Synthesis Lectures on Advances in Automotive Technology

This series covers the significant advances in new manufacturing techniques, low-cost sensors, high processing power, and ubiquitous real-time access to information that mean vehicles are rapidly changing and growing in complexity. These new technologies (including the inevitable evolution toward autonomous vehicles) will ultimately deliver substantial benefits to drivers, passengers and the environment. These publications cover the cutting edge of advanced automotive technologies.

Avesta Goodarzi · Yukun Lu ·
Amir Khajepour

Vehicle Suspension System Technology and Design

Second Edition

Springer

Avesta Goodarzi
Department of Mechanical
and Mechatronics Engineering
University of Waterloo
Waterloo, ON, Canada

Yukun Lu
Department of Mechanical
and Mechatronics Engineering
University of Waterloo
Waterloo, ON, Canada

Amir Khajepour
Department of Mechanical
and Mechatronics Engineering
University of Waterloo
Waterloo, ON, Canada

ISSN 2576-8107   ISSN 2576-8131   (electronic)
Synthesis Lectures on Advances in Automotive Technology
ISBN 978-3-031-21806-4   ISBN 978-3-031-21804-0   (eBook)
https://doi.org/10.1007/978-3-031-21804-0

This Springer imprint is published by the registered company Springer Nature Switzerland AG
The registered company address is: Gewerbestrasse 11, 6330 Cham, Switzerland

Preface

This book encompasses all essential aspects of suspension systems and provides an easy approach to their understanding and design. The book is intended specifically for undergraduate students and is accessible to anyone with an interest in learning about the foundations and design of suspension systems. This book uses a step-by-step approach using pictures, graphs, tables, and examples so that the reader may easily grasp difficult concepts.

After a short introduction, suspension systems and their components are discussed and reviewed in Chap. 1. The following chapter defines and examines suspension mechanisms and their geometrical features. In Chap. 3, suspension motions and ride models are derived to study vehicle ride comfort. This chapter ends with an analysis of suspension design factors and component sizing. Air suspension systems and their functionalities are reviewed and introduced in Chap. 4. The book ends with the development of adaptive suspension systems in Chap. 5.

Waterloo, Canada

Avesta Goodarzi
Yukun Lu
Amir Khajepour

Acknowledgments This book would not have been possible without the help of many people. We are particularly grateful to M. Naghibian, M. Ghare, C. Jarvis, S. M. Fard, and A. Pazooki, for their assistance in preparing the content of the book, and Ananya Chattoraj for editing and proofreading the book. We acknowledge the technical support from STAS (Belgium) and Shandong Meichen Industry Group Co. Ltd. (China). We are also thankful to Springer Nature for providing the publishing opportunity and for their consistent encouragement and support throughout this project.

Introduction

When people consider purchasing a new vehicle, they normally think of horsepower, torque, 0 to 100 km/h acceleration, and fuel economy. However, they are likely unaware of an important factor: the engine's power or vehicle speed is utterly useless if the driver cannot control the vehicle. Certainly, many people may recognize the importance of suspension for ride comfort, but they are less aware of the complete range of vehicle suspension duties. In fact, in addition to ride comfort, the suspension system plays an important role in vehicle performance, stability, and safety. Accordingly, automotive engineers turn their attention to the suspension system, an area usually ignored by customers considering a purchase.

Historic horse-drawn carts had an early form of a suspension system, where the platform swung on iron chains attached to a wheeled frame on the carriage (Fig. 1). This system was the basis for all suspension systems until the end of the nineteenth century. Obadiah Elliot is known as the first person that used a spring in the suspension system of a vehicle, and Mors of Paris first fitted a suspension system with shock absorbers in 1901.

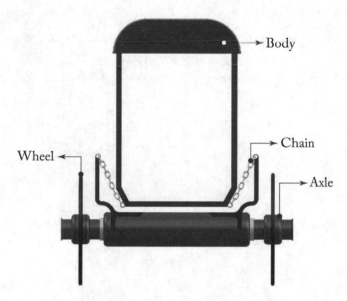

Fig. 1 Early form of suspension systems for horse-drawn carts

Today, there are many kinds of suspension systems with complex structures and elements, some of which will be discussed here.

This book begins with the definition of the suspension system and its function. It then goes on to describe the main components and desired features of suspension systems. This is followed by a classification of the different types of suspensions along with their advantages and disadvantages. Major suspension elements like springs, shock absorbers, and anti-roll bars are introduced. The design and analysis of a suspension mechanism along with its major parts are explained. There is also a section that reviews and discusses air suspension systems. Finally, the active and semi-active suspension systems, as well as some classic control techniques are introduced.

Contents

Suspensions Functions and Main Components

1

1.1 Suspension Systems

To many focusing on ride comfort, it may seem like the suspension system is merely a set of springs and shock absorbers that connect the wheels to the vehicle body. However, this is a very simplistic viewpoint of the suspension system. A vehicle suspension system provides a smooth ride over rough roads while ensuring that the wheels remain in contact with the ground and vehicle roll is minimized. The suspension system contains three major parts: a structure that supports the vehicle's weight and determines suspension geometry, a spring that converts kinematic energy to potential energy or vice versa, and a shock absorber that is a mechanical device designed to dissipate kinetic energy.

An automotive suspension connects a vehicle's wheels to its body while supporting the vehicle's weight. It allows for the relative motion between the wheel and vehicle body; theoretically, a suspension system should reduce a wheel's degree of freedom (DOF) from 6 to 2 on the rear axle and 3 on the front axle even though the suspension system must support propulsion, steering, brakes, and their associated forces. The relative motions of a wheel are its vertical movement, rotational movement about the lateral axes, and rotational movement about the vertical axes due to steering angle.

1.2 Functions of Suspension Systems

As previously mentioned, it is mostly assumed that the only function of a suspension system is the absorption of road roughness; however, the suspension of a vehicle needs to satisfy several requirements with partially conflicting aims as a result of different operating conditions. The suspension connects the vehicle's body to the ground, so all forces and moments between the two go through the suspension system. Thus, the suspension

© The Author(s), under exclusive license to Springer Nature Switzerland AG 2023
A. Goodarzi et al., *Vehicle Suspension System Technology and Design*,
Synthesis Lectures on Advances in Automotive Technology,
https://doi.org/10.1007/978-3-031-21804-0_1

system directly influences a vehicle's dynamic behavior. Automotive engineers usually study the functions of a suspension system through three important principles:

- **Ride Comfort:** Ride comfort is defined based on how a passenger feels within a moving vehicle. The most common duty of the suspension system is road isolation—isolating a vehicle body from road disturbances. Generally, ride quality can be quantified by the passenger compartment's level of vibration. There are a lot of inner and outer vibration sources in a vehicle. Inner vibration sources include the vehicle's engine and transmission, whereas road surface irregularities and aerodynamic forces are the outer vibration sources. The spectrum of vibration may be divided up according to ranges in frequency and classified as comfortable (0–25 Hz) or noisy and harsh (25–20,000 Hz).

- **Road Holding:** The forces on the contact point between a wheel and the road act on the vehicle body through the suspension system. The amount and direction of the forces determine the vehicle's behavior and performance. Therefore, one of the important tasks of the suspension system is road holding. The lateral and longitudinal forces generated by a tire depend directly on the normal tire force, which supports cornering, traction, and braking abilities. These terms are improved if the variation in the normal tire load is minimized. The other function of the suspension is supporting the vehicle's static weight. This task is performed well if the rattle space requirements in the vehicle are kept minimal.

- **Handling:** A good suspension system should ensure that the vehicle will be stable in every maneuver. However, perfect handling is more than stability. The vehicle should respond to the driver's inputs proportionally while smoothly following his/her steering/braking/accelerating commands. The vehicle behavior must be predictable, and behavioral information should accordingly be communicated to the driver. Suspension systems can affect vehicle handling in many ways: they can minimize the vehicle's roll and pitch motion, control the wheels' angles, and decrease the lateral load transfer during cornering.

1.3 Main Components of Suspension Systems

A vehicle suspension system is made of four main components: mechanism, spring, shock absorber, and bushings as shown in Fig. 1.1.

- **Mechanism:** The suspension mechanism might contain one or several arms that connect a wheel to the vehicle body. They transfer all forces and moments in different directions between the vehicle body and the ground. The mechanism determines some of the most important characteristics of a suspension system. It determines the

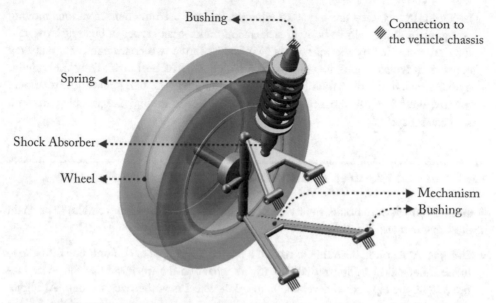

Fig. 1.1 The main components of a suspension system

suspension geometry and wheel angles and their relative motions. Variation in wheel angles during suspension travel causes a change in tire forces, which affects the vehicle's road holding and handling. The main weight of a suspension system arises from its mechanism. Using heavy materials in its construction decreases the ride quality; whereas, light materials, although improve ride quality, are more expensive.

- **Spring:** The spring is usually a winding wire or several strips of metal that have elastic properties. It supports the vehicle's weight and makes a suspension tolerable for passengers. To best understand suspension behavior, the most important component requiring study is the spring. However, with its importance, a conflict arises: when using high stiffness springs, the vehicle exhibits good road holding and handling but with a noticeably decreased ride comfort. This creates a condition of limitation when choosing an appropriate spring stiffness. The spring weight and size may also make this accommodation difficult.
- **Shock absorber:** The shock absorber is a mechanical or hydraulic device to dampen impulses. A high damping shock absorber compromises the vehicle's ride quality to immediately dampen impulses to improve handling and road holding.
- **Bushings:** The bushings prevent the direct contact of two metal objects to isolate noise and minimize vibration. Soft materials such as rubber are used in bushings for

isolation. In fact, they are a type of vibration isolator used to connect various moving components to the vehicle body or suspension frame. Many types of bushing exist, and they are classified by the number of DOF between the two connected parts that they support. Revolute joints are the most common type of bushings. They are annular cylindrical and support a rotational relative motion, whereas, ball joints allow rotational relative motion in all directions. Bushings are some of the most expensive parts in a suspension system.

1.4 Desired Features of Suspension Systems

A suspension system should satisfy certain requirements for use in vehicles. The main desired features are as follows:

- **Independency:** It is desirable to have the movement of a wheel on one side of the axle to be independent of the movement of the wheel on the other side of the axle. Figure 1.2 on the top shows a vehicle whose left wheel is going over a bump. At higher speeds, the wheel can negotiate the bump without disturbing the other wheel. This is only possible when each wheel has an independent suspension. Independency of wheel movement improves a vehicle's ride comfort, road holding, and handling.
- **Good camber control:** The camber angle is the wheel angle about its longitudinal axis (this will be comprehensively discussed in Sect. 3.2.1). A negative camber is desired since it results in improved handling, however, the convex shape of roads tends toward a positive camber to reduce tire wear. Due to road bump and body roll, the camber will ultimately change. Using a well-designed suspension geometry, we can control the camber angle (Figs. 1.3 and 1.4).
- **Good body roll control:** Each suspension system has a roll center. The hypostatic line that connects the front and rear suspensions' roll centers is called the roll axis. The vehicle's body rolls about this line during cornering maneuvers. It is necessary to analyze the roll axis because of its effect on the body roll and lateral vehicle behavior. The design of the suspension geometry should account for the best location of the roll axis to optimize vehicle body roll motion.
- **Good space efficiency:** It comes as no surprise that the space utilized by a suspension system may create difficulties for the installation of other components of the vehicle. Under the hood, the suspension system should leave enough room for an engine and other components. Also, the suspension of the rear axle should not interfere with the vehicle trunk and instead, occupy only its internal space (Fig. 1.5).
- **Good structural efficiency:** The suspension system should be able to handle the vehicle's weight and all applied forces and moments in the contact area between the wheel and the road. The suspension mechanism must feed loads into the body in a well-distributed manner and prevent the transfer of concentrated forces onto the vehicle's body (Fig. 1.6).

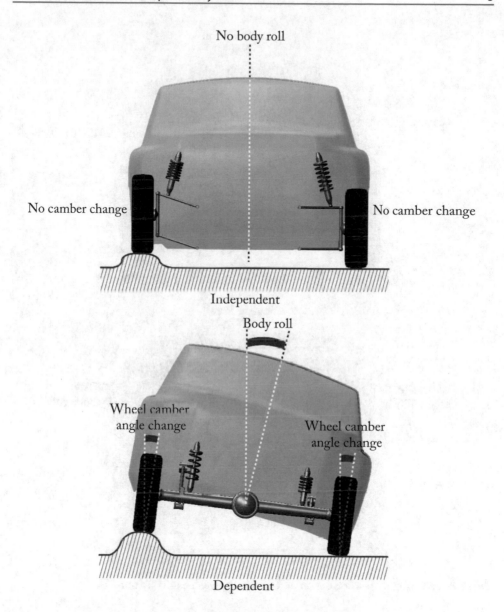

No body roll

No camber change   No camber change

Independent

Body roll

Wheel camber
angle change   Wheel camber
                                    angle change

Dependent

Fig. 1.2 Independent versus dependent suspension systems

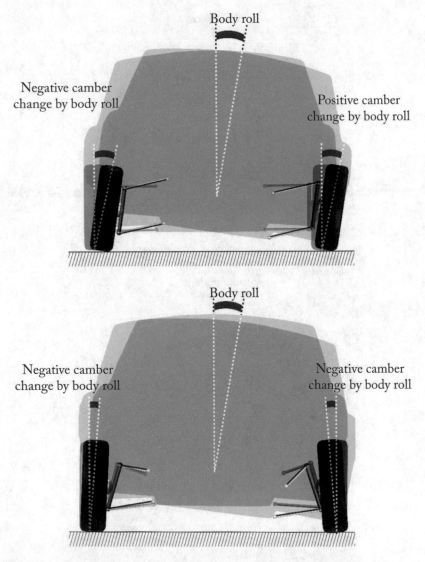

Fig. 1.3 Effects of suspension design on Camber change due to body roll

- **Good isolation:** Improving ride quality and isolating road roughness is one of the most important tasks of a suspension system.
- **Low weight:** Due to road irregularities, the kinematic energy of a suspension system is proportional to its mass. Higher kinematic energy results in stronger transmitted shocks to the vehicle body. This effect clearly decreases the ride quality. To minimize this negative effect, we should minimize the suspension mass by using optimized designs

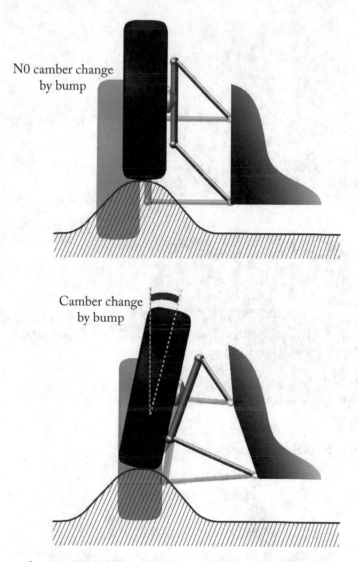

Fig. 1.4 Effects of suspension design on Camber change due to road bump

and/or lightweight material. Lightweight materials may increase the cost and therefore, a balanced design is needed for any suspension system.

- **Long life:** No one enjoys having to repair their car frequently, so the suspension must be as durable as any other part of a car. A durable system can resist wear, pressure, or damage, all of which play important roles in the success of a product.
- **Low cost:** While defining a low enough cost is a subjective matter, the suspension as a vehicle sub-system should be affordable. High-performance suspension systems are more expensive and mainly used in premium vehicles. Using a high number of

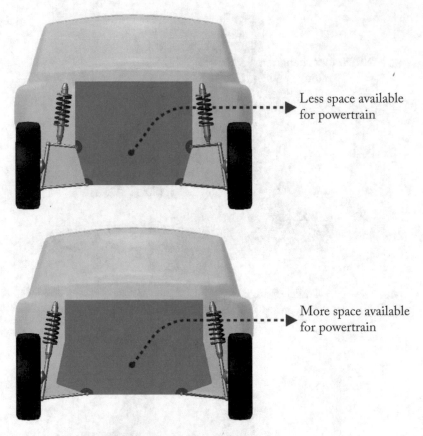

Less space available
for powertrain

More space available
for powertrain

Fig. 1.5 Space efficiency of suspension systems

Fig. 1.6 Structural efficiency in suspension systems

bushings and lightweight materials certainly improves the ride quality, noise isolation, and performance of the system, but they also increase the cost of the product.

- **Others:** Other suspension features may include anti-dives and anti-squats. When a vehicle is braking, a dive occurs, where the front of the vehicle dips and the tail rises. A similar but opposite action, a squat, happens during acceleration. This rotational movement is slight but since the human body is very sensitive to pitch motion, mitigating this movement in the passenger cabin allows for greater ride quality.

Classification of Suspension Systems

2

2.1 Different Suspension Types

2.1.1 Solid Axle

A solid axle is known to be the first suspension type used in vehicles where the wheels are mounted at either ends of a rigid beam (Fig. 2.1). A solid axle system generally uses leaf springs, however, there are some equipped with coil springs. A solid axle suspension system is not classified as an independent type of suspension, so any movement of one wheel is transmitted to the other wheel. Solid axles have some advantages in their load-carrying capacity; solid axles are also affordable and durable. However, they have many disadvantages when considering passenger cars. Leaf springs result in unwanted transmitted noise and vibration. Rigid beams considerably reduce ride quality, and their size and dimensions demand enough space above the axle to accommodate the springs. Solid axle suspensions are commonly used in commercial vehicles where a high load carrying capacity is required. Table 2.1 presents a summary of details regarding the solid axle.

2.1.2 Torsion Beam

The torsion beam is known as the most popular rear axle suspension system for small and medium size, front-wheel drive vehicles. A torsion beam suspension consists of two trailing arms with a flexible cross member in between, shown in Fig. 2.2. The cross member absorbs all forces and moments and simultaneously works as an anti-roll bar.

The torsion beam is a semi-independent suspension. It boasts few components and a simple structure; therefore, it is light, cheap, easily assembled, and requires little space. The lightweight nature of the system results in improved ride comfort, but cross-member deformation lends itself to the tendency of lateral force oversteering and poor handling.

© The Author(s), under exclusive license to Springer Nature Switzerland AG 2023   11
A. Goodarzi et al., *Vehicle Suspension System Technology and Design*,
Synthesis Lectures on Advances in Automotive Technology,
https://doi.org/10.1007/978-3-031-21804-0_2

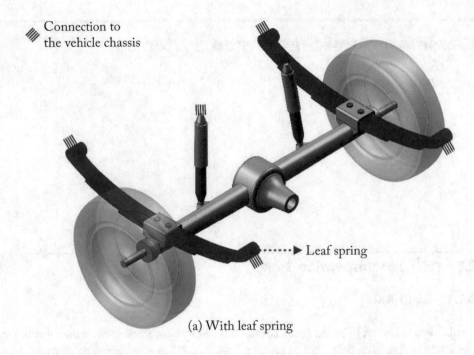

Connection to
the vehicle chassis

Leaf spring

(a) With leaf spring

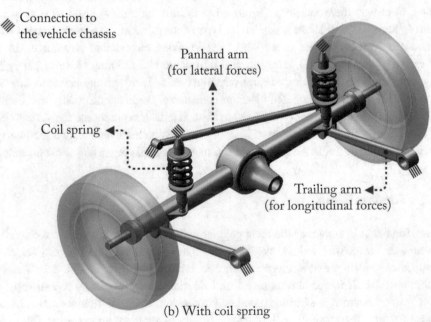

Connection to
the vehicle chassis

Panhard arm
(for lateral forces)

Coil spring

Trailing arm
(for longitudinal forces)

(b) With coil spring

Fig. 2.1 Different types of solid axle suspension

Table 2.1 Advantages and disadvantages of a solid axle suspension

Features	Description	Verdict
Independency	Mutual wheel influence	Bad
Camber angle	Fixed wheels to the axle	Bad
Roll angle	High roll center height	Moderate
Space efficiency	Need more space	Bad
Structural efficiency	Concentrated forces at mounting points	Bad
Isolation	Noisy due to leaf springs motion against each other	Bad
Weight	Heavy axle	Bad
Lifespan	Less moving components and maintenance	Good
Cost	Fewer components and no complex parts	Good

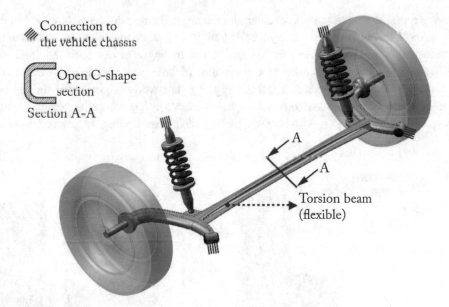

Fig. 2.2 Torsion beam suspension

Furthermore, cross-member flexibility diminishes camber and body roll control. A summary of torsion beam axles is presented in Table 2.2.

2.1.3 Trailing Arm

The trailing arm suspension is used in the rear axle with either front or rear-wheel drive vehicles, and it is classified as an independent suspension system. This suspension system consists of a control arm that is longitudinally connected to the body, while the wheels

Table 2.2 Advantages and disadvantages of a torsion beam suspension

Features	Description	Verdict
Independency	Semi-independent	Moderate
Camber angle	More flexibility due to cross member	Moderate
Roll angle	More flexibility due to cross member	Moderate
Space efficiency	Needs smaller space	Good
Structural efficiency	Concentrated forces in few mounting points	Bad
Isolation	Better isolation and no leaf spring noise	Moderate
Weight	Simple systems with light components	Good
Lifespan	Less moving components and maintenance	Good
Cost	Fewer components and no complex parts	Good

remain parallel to the body, so no camber and toe angle changes are caused by the wheel's vertical movement. However, during a body roll, the wheel camber changes significantly. To reduce this unwanted effect, the design of the trailing arm has been modified to a semi-trailing arm suspension, where the revolt axis of the control arm makes an angle with the vehicle's longitudinal axis as shown in Fig. 2.3. In a well-designed semi-trailing arm suspension, minor camber and toe angle changes happen either due to the wheel's vertical movement or the body roll. Although trailing and semi-trailing suspension systems

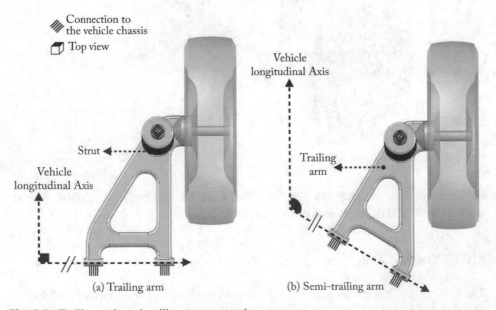

Fig. 2.3 Trailing and semi-trailing arm suspension systems

provide good ride comfort, durability, and handling performance, they are now out-of-date because there are more effective independent suspension systems for vehicles. Figure 2.3 shows the trailing and semi-trailing arm suspensions. Their main features are listed in Table 2.3.

2.1.4 Double Wishbone

The double wishbone suspension is used in front and rear axles, and it consists of two control arms to hold the wheel as shown in Fig. 2.4. The main advantage of the double wishbone suspension is its customizable kinematic design. The positions, lengths, and angles of the control arms specify the roll and pitch behavior of the vehicle (through the roll center height and the pitch pole position), and the double wishbone can influence wheel angle changes after bump-related movement. The designer can adjust the suspension characteristics for the particular needs of different vehicle types to achieve the best handling and stability performances.

The double wishbone suspension is large laterally, so it minimizes the undesirable effects on the ride quality. This type of suspension is relatively expensive due to its complexity and number of joints. Connecting the spring to the lower wishbone as shown in Fig. 2.4a, or to the upper wishbone as shown in Fig. 2.4b, may cause interference with the driving shaft or extra space for the suspension packaging, respectively. A summary of the features of double wishbone suspensions is listed in Table 2.4.

2.1.5 Macpherson Strut

The Macpherson strut is usually used as the front axle of small and medium size front wheel drive vehicles. The upper control arm in the double wishbone is replaced by a strut that is rigidly connected to the wheel's spindle (Fig. 2.5). The lower control arm mainly

Table 2.3 Advantages and disadvantages of the trailing/semi-trailing arm suspension

Features	Description	Verdict
Independency	Fully independent	Excellent
Camber angle	By optimum design of the revolt axis	Good
Roll angle	By optimum design of the revolt axis	Good
Space efficiency	Needs smaller space	Good
Structural efficiency	Two connection points per wheel	Moderate
Isolation	Bushings are used at the connection points	Good
Weight	Dependent on the control arm design	Good
Lifespan	No complex elements	Good
Cost	Needs four bushings per axle	Moderate

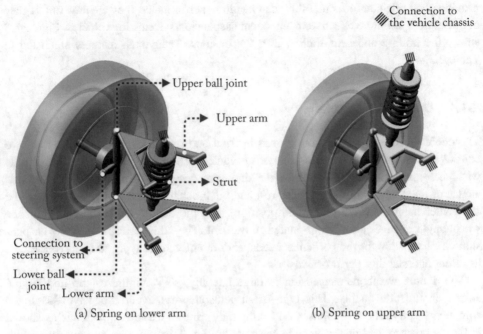

(a) Spring on lower arm   (b) Spring on upper arm

Fig. 2.4 Different types of double wishbone suspension

Table 2.4 Advantages and disadvantages of a double wishbone suspension system

Features	Description	Verdict
Independency	Fully independent	Excellent
Camber angle	Customizable	Good
Roll angle	Customizable	Good
Space efficiency	Needs larger space, especially in the lateral direction	Bad
Structural efficiency	Four connection points per wheel	Good
Isolation	Bushings are used at the connection points	Good
Weight	Dependent on wishbone design and materials	Moderate
Lifespan	No complex elements	Good
Cost	Could be expensive	Moderate

bears lateral and longitudinal forces, and its structure and shape play important roles in wheel compliance control. This independent suspension is small enough to allow for an under-the-hood installation of the engine following the installation of the suspension. It is cheaper and lighter than the double wishbone suspension; however, replacing the upper arm with the strut decreases the roll, pitch, and wheel angle control, so from a vehicle handling point of view, the Macpherson strut is merely acceptable. A summary of the features of double wishbone suspensions is listed in Table 2.5.

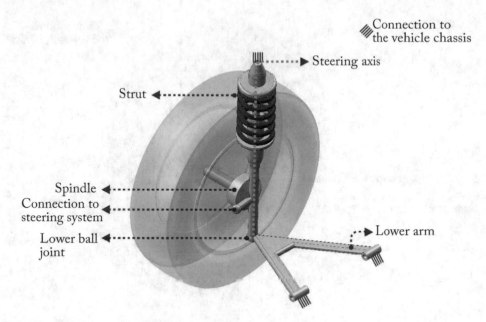

Connection to the vehicle chassis

Steering axis

Strut

Spindle

Connection to steering system

Lower ball joint

Lower arm

Fig. 2.5 Macpherson strut suspension system

Table 2.5 Advantages and disadvantages of a Macpherson strut suspension

Features	Description	Verdict
Independency	Fully independent	Excellent
Camber angle	Not as good as Double Wishbone	Moderate
Roll angle	Not as good as Double Wishbone	Moderate
Space efficiency	Needs smaller space	Good
Structural efficiency	Concentrated forces on three connection points	Moderate
Isolation	Use of bushings at the connection points	Good
Weight	Lightweight	Good
Lifespan	Strut has a more complex design than other suspensions	Moderate
Cost	Simpler design compared to Double Wishbone	Good

2.1.6 Multi-link

In recent years, multi-link suspension systems have become popular in the rear axle of luxury vehicles. As shown in Fig. 2.6, a multi-link suspension is formed with five or more links to make a truss structure, where most of the links are just under the axial load. There are many design parameters in this configuration, so multi-link suspensions can be designed to better control the wheel angles and body movements. Also, the truss structure of a multi-link suspension makes it very light. The high number of links and joints make multi-link suspensions complicated and expensive with more maintenance costs. A summary of the features of multi-link suspensions is listed in Table 2.6.

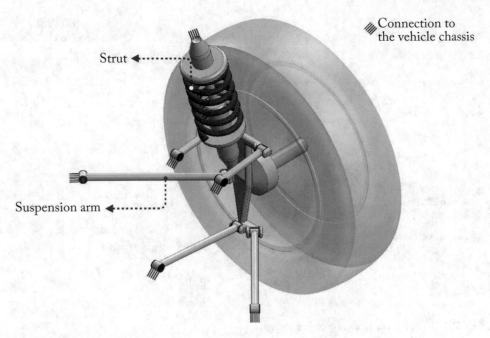

Fig. 2.6 Multi-link suspension system

Table 2.6 Advantages and disadvantages of a multi-link suspension system

Features	Description	Verdict
Independency	Fully independent	Excellent
Camber angle	Can be optimized through the design	Excellent
Roll angle	Can be optimized through the design	Excellent
Space efficiency	Can be designed for tight spaces	Moderate
Structural efficiency	Several connection points	Excellent
Isolation	Use of bushings at the connection points	Good
Weight	Truss structure	Excellent
Lifespan	Needs extra maintenance due to its complexity	Moderate
Cost	Expensive	Bad

Example 2.1: Select the appropriate suspension system type in the front and rear axles of each of the following vehicles:

(a) A compact sedan
(b) A mid-size SUV
(c) A commercial van
(d) A full-size luxury sedan

Answers

(a) Compact sedans are considered to be affordable vehicles, and a space-efficient suspension system is crucial due to their compact nature. Therefore, an appropriate suspension system should be relatively small and inexpensive. As such, the most common configuration in the compact sedan is the Macpherson strut system for the front axle, and the torsion beam system for the rear axle. This configuration provides a moderate ride quality and handling at a cost-effective range.

(b) A mid-size SUV may sometimes drive off-road, where significant forces act on the suspension system. Thus, durability, structural efficiency, and isolation should be prioritized. From a safety perspective, body roll control is paramount since SUVs are susceptible to roll over due to their high height. Therefore, for a mid-size SUV, an improved Macpherson strut or a double wishbone system should be implemented in the front axle, and a multi-link system should be chosen for the rear axle. The shape and number of bushings of the Macpherson strut are improved when designed for an SUV due to greater load-bearing capacity

(c) A commercial van is usually used to distribute goods in cities, so the suspension system in the rear axle must be able to handle large loads every day. Durability and low maintenance cost are very important in this type of vehicle. An improved Macpherson strut or a double wishbone system is used in the front axle, and a solid axle is suitable for the rear axle.

(d) Consumers purchasing a full-size luxury sedan expect the very best in terms of vehicle features and performances. The suspension system should provide excellent handling through accurate camber and roll control, and superb ride comfort quality through perfect vibration isolation. The double wishbone or multi-link system can be used in the front axle of a full-size luxury sedan, and a multi-link can be used in the rear axle.

2.2   Spring Types

A spring is an elastic object used to convert kinematic energy into potential energy and vice versa. The spring is an important component in a suspension system as it bears the normal forces and is crucial to vehicle ride quality and handling. Therefore, different types of springs were invented to improve various suspension performances.

2.2.1   Coil Springs

Coil springs appeared in the early fifteenth century. Coil springs are made of elastic materials shaped into a helix. The traditional coil spring as shown in Fig. 2.7a obeys Hooke's law and has a constant spring rate; however, in modern cars, it should obey a

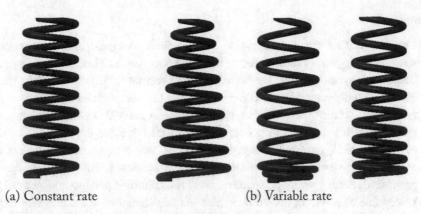

(a) Constant rate   (b) Variable rate

Fig. 2.7 Coil springs with **a** constant rate and **b** variable rate

variable spring rate since the spring provides the vehicle with a soft ride when traveling with light loads and a firmer ride when traveling with heavier loads. Also, this variable spring rate provides improved body and angle control. There are multiple methods of creating the variable rate coil spring as described below and shown in Fig. 2.7b.

- Variable diameter winding of the elastic material which decreases toward the ends of spring.
- Variable bar diameter where the bar tapers nearing each end.
- An extended elastomer bump stop.
- Conical or beehive shape winding of the elastic material.

2.2.2   Torsion Bars

A torsion bar is a torsional spring used as the main spring in a suspension system. It is constructed as a long bar whose one end is connected to the vehicle chassis and its other end is joined to the suspension control arm as shown in Fig. 2.8. The vertical motion of the wheel results in a rotational movement in the bar around its axis, which is resisted by the bar's torsional stiffness. Rear suspension torsion bars are mainly used in small European cars, whereas, some sport utility vehicles and trucks have them in their front suspensions.

2.2.3   Leaf Springs

Leaf springs consist of several arc-shaped longitudinal strips, as shown in Fig. 2.9, where the center of the arc is supported by the axle. Usually, one end of the leaf spring is attached directly to the frame and the other end is connected through a shackle link. Leaf springs are mainly used on solid rear axles, which mainly appear in trucks.

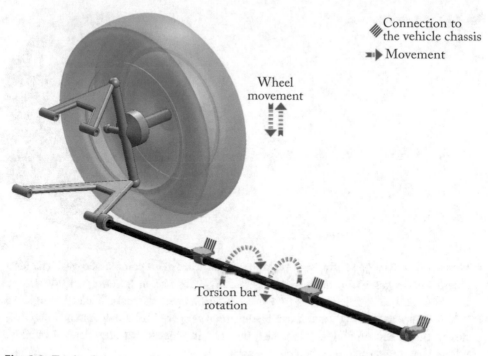

Connection to
the vehicle chassis

Movement

Wheel
movement

Torsion bar
rotation

Fig. 2.8 Torsion bar

Fig. 2.9 Leaf spring

2.2.4 Alternative Springs

Alternative springs are used in advanced suspension systems such as the self-leveling and hydro-pneumatic systems. The main advantage of this kind of spring is the independency of suspension characteristics from carrying loads. Alternative springs include the air spring and the gas spring.

- **Air spring:** An air spring, shown in Fig. 2.10, is a pillow that inflates via an electric air pump. It provides high-quality ride comfort without dependency on carrying loads. The cost of an air spring is high, and they are mainly used in luxury sedans and heavy vehicles like buses and trucks. Air springs provide a self-leveling feature that is useful in many vehicles.

Fig. 2.10 Air spring

- **Gas spring:** The gas spring is a piston-cylinder filled with compressed gas. The strut shape of the gas spring makes it very practical for use in a variety of suspension systems. A gas spring can also be used as a damper; as such, it can be called a "slow-dampened spring", and it can be designed for parts that are operated slowly like heavy doors and windows. It can also be used in objects that are operated quickly, where they are termed as a "quick gas spring."

2.2.5 Advanced Materials Springs

Composite materials have the characteristics that can replace metal springs. Composite springs are lightweight, durable, and highly corrosion resistant. However, their difficult manufacturing process and high cost prevent them from mass implementation in all vehicle classes.

2.3 Shock Absorber Types

A shock absorber, also known as a damper in the automotive industry, has a piston that moves inside a sealed, oil-filled cylinder. There are one-way valves on the piston that allow the oil to slowly flow from one chamber to another. In a running vehicle, shock absorbers dampen vibrations to control vehicle handling and ride comfort. Usually, shock absorbers convert kinematic energy to heat. Three types of shock absorbers are described below.

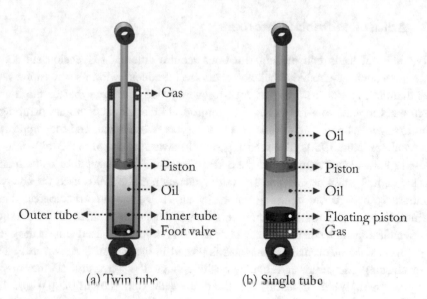

Fig. 2.11 Shock absorbers **a** Twin tube **b** Mono tube

2.3.1 Twin-tube

The twin-tube shock absorber consists of two cylindrical tubes as shown in Fig. 2.11a. At the bottom of the inner tube, there is a foot valve that allows oil to flow between the tubes. The piston moves up and down into the inner tube while the outer tube serves as a reservoir. Oil moves between the upper and lower chambers of the inner tube via the piston valve as well as the inner and outer tubes via the foot valve. The orifice valves in the piston control most of the damping with assistance from the foot valve.

2.3.2 Mono-tube

As its name indicates, a mono-tube shock absorber consists of one tube that is divided into sections by two pistons, the working piston, and the floating piston, as shown in Fig. 2.11b. The gas and oil are separated by the floating piston, which plays a role analogous to that of the foot valve in twin-tube shock absorbers. Similar to that of the twin-tube, the orifices in the piston valve produce most of the damping, and during more aggressive movements, the floating piston pushes farther into the gas chamber to quickly increase gas pressure and provide an additional damping force. The length of the cylinder and piston in a mono-tube is consistently larger than those in the twin-tube since they require a larger area to operate.

2.3.3 Adjustable Shock Absorbers

Ideally, different loads and driving situations require different suspension stiffness and damping. As such, adjustable shock absorbers are introduced within suspension systems. Lower damping results in improved high-frequency ride comfort, whereas the improvement in handling and low-frequency ride comfort are the results of higher damping.

The first generation of adjustable shock absorbers was manual, and they appeared on luxury vehicles in the 1950s. The second generation was introduced in the mid-1980s, and they used a solenoid valve to control the hydraulic circuit. They were able to automatically adjust the damping at 3 or 4 levels. The third generation in the 1990s used variable valves to continually change the damping value. In the 2000s, magneto-rheological dampers were introduced as the fourth generation, where the cylinder is filled with a mixture of magnetized iron particles in a synthetic hydrocarbon oil. A coil is used around the cylinder body while a stake of permanent magnets is placed in the piston as shown in Fig. 2.12. The electromagnetic field can continuously change the viscosity of the magneto-rheological fluid. When it is off, the fluid flows through the piston passage freely. Whereas, when the magnetic field increases, the fluid becomes more viscous, which changes the damping force.

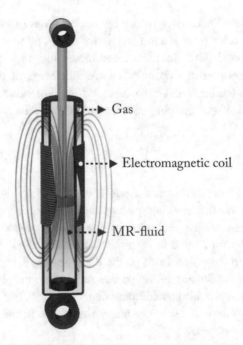

Fig. 2.12 Magneto-rheological shock absorber

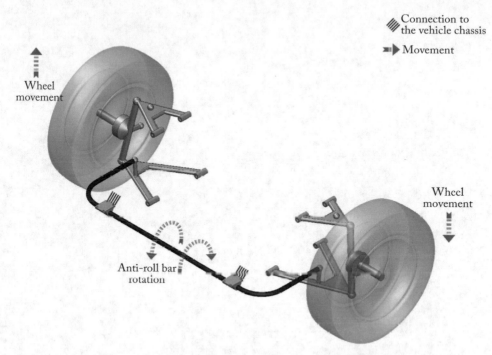

Fig. 2.13 Anti-roll bar connects the wheels of an axle to provide torsional stiffness

2.3.4   Anti-roll Bar

As shown in Fig. 2.13, to reduce body roll during cornering, an anti-roll bar connects the two opposite wheels on an axle through a torsional bar that acts as a torsional spring. When both wheels of an axle move simultaneously in the same direction, the anti-roll bar remains inactive, but if the wheels move in opposite directions, the bar is subjected to torsion and is forced to twist. The anti-roll bar decreases body roll by increasing roll stiffness without affecting the main suspension stiffness.

Analysis and Design of Suspension Mechanisms

3

3.1 Suspension Geometry

Suspension geometry plays an important role in wheel angles and movements, all of which affect vehicle road holding, handling, and safety. Suspension kinematic movements include wheel movements during vertical travel and steering as well as the movements due to the body roll and pitch. To study suspension kinematics, the SAE vehicle's coordinate system X–Y–Z at the vehicle's center of gravity, and the wheel's coordinate system x–y–z are defined in Fig. 3.1.

3.2 Roll Center and Roll Axis

The roll center is a point on the y–z and X–Z planes where the lateral force applics to the vehicle body without kinematic roll occurring. The axis in the X–Z plane that goes through the front and rear roll centers is called the roll axis, and this is the axis that the body rolls around. The heights of the roll centers and roll axis play an important role in a vehicle's handling performance and ride quality. To determine the roll center, Arnold Kennedy's theorem is used. If three bodies have motion relative to each other, their instantaneous centers should lie in a straight line.

To find the roll center in a double wishbone suspension, only the directions of the control arms are important. The virtual center of rotation, P, is obtained by extending the upper and lower control arms. Instantaneously, P is linked with the center of the tire and extended to find the point where it intersects with the X–Z plane; this point is called the roll center. Figure 3.2 shows the roll center determination process in a double wishbone system.

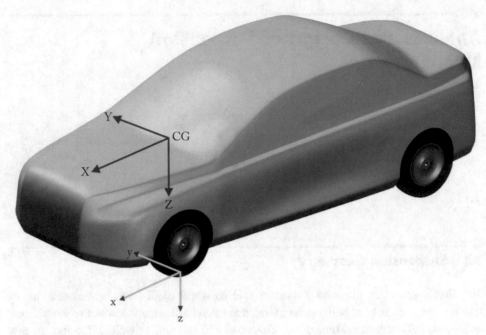

Fig. 3.1 Vehicle and wheel coordinate systems

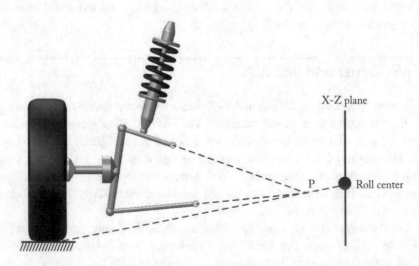

Fig. 3.2 Body roll center for a double wishbone suspension

In Macpherson suspensions, a vertical axis is created in the strut's end point where the intersection of it and the extended lower control arm provides a virtual center of rotation, P. Figure 3.3 shows the roll center determination process on Macpherson systems.

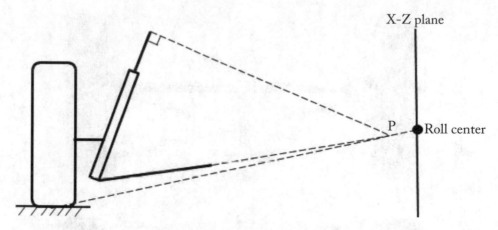

X-Z plane

P ● Roll center

Fig. 3.3 Body roll center for a Macpherson strut suspension

In the torsion beam axle, whose top view is shown in Fig. 3.4a, the bearing point is connected to the torsion beam center point, s, and intersects the y–z plane. From the rear view in Fig. 3.4b, the virtual center of rotation, P, is linked with the center of the tire, and its intersection on the X–Z plane provides the roll center.

3.2.1   Camber Angle

The camber angle is the angle between the x–z plane of the wheel and a line perpendicular to the plane of the road on the Y–Z plane of the vehicle. This angle is negative if the wheel is inclined inward as shown in Fig. 3.5a and positive when inclined outward as shown in Fig. 3.5b. A positive camber would be useful for a reduction in tire wear due to the road profile. On the other hand, a negative camber provides better tire grip and improves handling.

The initial camber is set for a static wheel with factory settings where the rear wheels are usually more negative in camber angle than the front wheels. Wheel travel changes the camber angle as shown in Fig. 3.6. Proper suspension design should result in a negative camber during wheel travel bump and a positive camber during rebound. This design is useful during cornering because the inner wheel that travels during a bump supports bigger loads than the outer wheel in the rebound.

3.2.2   Toe Angle

The toe is the angle between the x–z plane of the wheel and the X–Z plane of the vehicle. A "toe-in" occurs when the front part of the wheel is turned inward, as shown in Fig. 3.7, and a "toe-out" occurs when the wheel is turned outwards. Rolling resistance and braking

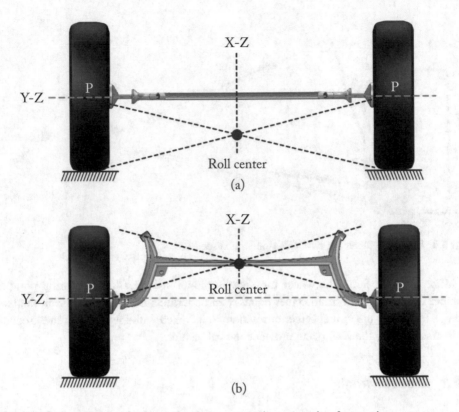

Fig. 3.4 Body roll center in the torsion beam suspension **a** top view **b** rear view

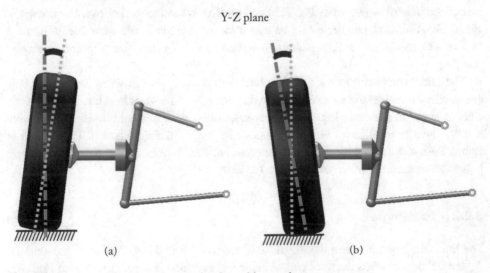

Fig. 3.5 Camber angle **a** negative camber **b** positive angle

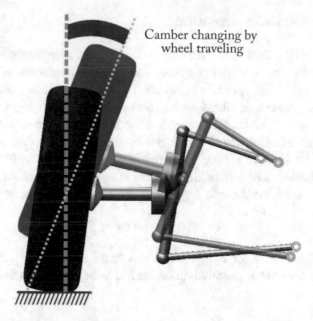

Fig. 3.6 Camber changing by wheel travel

Fig. 3.7 Toe-in in front wheels

forces push the wheel slightly backward resulting in toe-out situations, and when traction pulls the wheel forward, it results in a toe-in situation. The toe angle should ideally be zero for the lowest amount of tire wear and rolling resistance. Therefore, manufacturers tend to define a toe angle on a static wheel that goes to zero under working conditions. Usually, to offset the braking forces on the front axle that normally results in toe-out situations, manufacturers set the front axle with toe-in angles. This is done because the vehicle may be unstable if the absolute value of the angle presents as a toe-out. A rear drive axle is set with a toe-out because of its traction force, and a rear non-drive axle adjusts to toe-in to compensate for rolling resistance.

3.3   Fundamentals of Vibration

Vibration theory is the field of science that studies the oscillatory motions of bodies and their associated forces. A vibrating system needs two elements: mass and elasticity; therefore, all un-rigid bodies are able to vibrate. The vibration system would have a shock absorber that turns kinematic vibrations into heat resulting in the vibrations diminishing. There are two types of vibration: free vibration where there is no external force to the system and forced vibration where at least one external force is applied to the body. The undamped time response of a body under free vibration is shown in Fig. 3.8; the body repeats its oscillation about an equilibrium point at equal intervals of time.

The equation of the motion of a system that vibrates freely can be written as:

$$x(t) = A \sin \omega t \tag{3.1}$$

where, A is the amplitude of oscillation, which is the maximum displacement of the oscillatory mass from the equilibrium point, and ω is the natural frequency that can be defined as:

$$\omega = \frac{2\pi}{T} = 2\pi f \tag{3.2}$$

In Eq. (3.2), T is the period of oscillation, which is the duration of time that a system needs to vibrate one cycle, and f is the frequency. The relationship between period and frequency is:

$$T = \frac{1}{f} \tag{3.3}$$

Fig. 3.8 A simple sinusoidal vibration of a system under free vibration

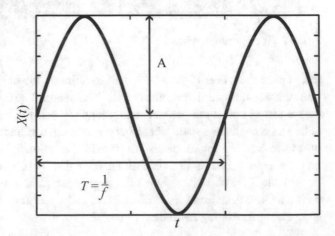

Frequency is the number of cycles per unit time. Oscillations can usually be classified according to their frequency. The lower the oscillation frequency, the longer its wavelength and the lower its energy. A high-frequency oscillation has a shorter wavelength but higher energy.

3.3.1 Vehicle Noise, Vibration, and Harshness

A car has mass and elasticity, so according to the definition of vibration, it can have both free and forced vibrations. There are a lot of rotational parts in a vehicle such as the engine, gearbox, wheels, differential, etc. Each of them can generate forces that are applied to the vehicle's body. Not only do they have rotational motion, but they have also been installed asymmetrically within the vehicle, which makes vibrational situations worse. Moreover, regardless of the vehicle segment, a driver drives the car on different road conditions cars can be driven on-roads or off-roads. In addition, different sounds are also generated because of the contact between the tire and the ground, air flow over the vehicle body, engine, etc. Such vibrations can considerably harm the vehicle ride experience.

Based on the aforementioned details, a car is exposed to a wide range of oscillations with different frequencies. The source of these vibrations can be internal or external. According to the vibration spectrum, we can classify them into three groups: Noise, Vibration, and Harshness (NVH). These classifications are based on the frequency of vibration, and each term refers to a specific frequency range that creates a variety of effects on the vehicle and passengers. These terms are defined as:

- **Vibration:** Oscillations with a frequency less than 25 Hz are known as vibrations that are associated with the oscillatory motion of the vehicle body; not only can this motion be seen, but it can also easily be felt by passengers. This sort of vibration is tangible, and passengers usually judge a vehicle's ride according to this range of vibrations. Thus, automotive engineers should pay close attention to control vibrations especially in this range. As mentioned earlier, one of the crucial tasks of a suspension system is to mitigate vibrations resulted from the contact between the tires and the ground.
- **Harshness:** Oscillations with a frequency higher than 25 Hz and less than 100 Hz are called harshness. It cannot be seen, but passengers can feel the discomfort caused by this unwanted vibration. Harshness can be felt as shaking seats and dashboard; passengers cannot evaluate the harshness of a vehicle, but during long drives, they feel more tired after being in a vehicle with high harshness. Harshness can be generated by both external sources like road profile, and internal sources such as the engine, transmission, and brake.
- **Noise:** Oscillations with a frequency higher than 100 Hz are classified as noise, and this refers to an unwanted sound. These oscillations cannot be seen or felt, but they can be heard by passengers. Passengers can easily hear and be disturbed by unwanted

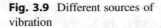 **Fig. 3.9** Different sources of vibration

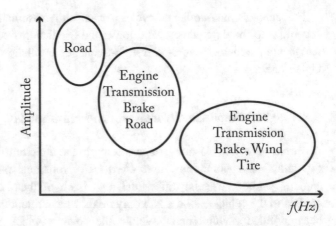

sounds. Noise can be produced by internal components like the engine, transmission, and brakes, which are audible even at low speeds, or it can also be produced by external sources such as wind and tires, which can only be heard at high speeds.

Figure 3.9 shows the sources of vibration, harshness, and noise; moreover, it also shows the typical relationship between amplitude and frequency for each group. It is obvious that whenever the frequency increases, the amplitude decreases, and vice versa.

3.3.2 Road Profile

As mentioned, one of the crucial tasks of the suspension is to cope with the vibrations resulting from the contact between the tire and the road. In order to design an effective suspension, we need to comprehensively know the road. Therefore, we will review the sinusoidal road profile, which is the simplest form of road, and a study of the real road profile will follow. Finally, some typical road profiles to test the suspension will be introduced.

a. Sinusoidal Road Profile

The road's roughness is the principal source of vibration in a vehicle. In the other words, road irregularities can induce oscillatory bounce, pitch, and roll motions on a sprung mass shown in Fig. 3.10. The simplest form of road roughness is in a sinusoidal shape. The vibrations transmit directly onto the vehicle and negatively affect ride comfort, which can result in increased passenger and driver tiredness and discomfort. Moreover, a vehicle's unsprung mass or wheel is also excited by the road, which affects the tires' holding forces, and can eventually disturb the handling, traction, and braking performance of the vehicle.

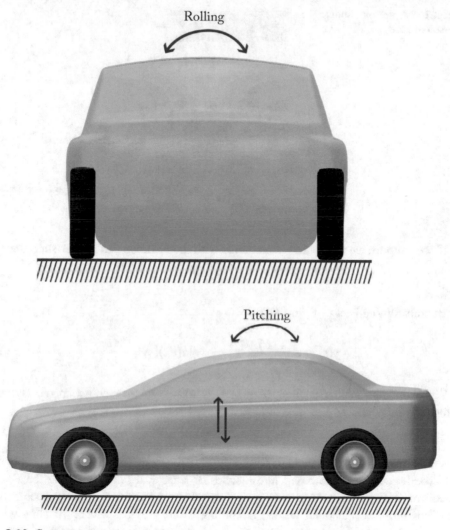

Fig. 3.10 Sprung mass movement

A rolling tire can track along the ideal road profile, which is considered to be a simple sinusoidal shape shown in Fig. 3.11. The road profile can be presented with the following basic formulation:

$$y(x) = d \sin\left(\frac{2\pi}{\lambda}x\right) \tag{3.4}$$

Fig. 3.11 A tire on a simple sinusoidal road

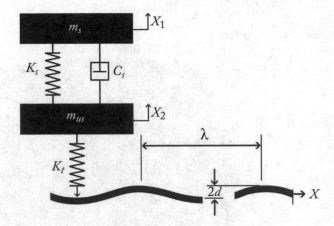

If we suppose the tire remains in contact with the road, we can further suppose:

$$x = ut \tag{3.5}$$

By substituting Eqs. (3.5) in (3.4):

$$y(t) = d \sin\left(\frac{2\pi u}{\lambda} t\right) = d \sin(\omega t), \omega = \frac{2\pi u}{\lambda} \tag{3.6}$$

Therefore, the road excitation frequency is related to the vehicle speed, and it is converse to the road profile wavelength.

b. **Real Road Profile**

It is too unrealistic to think that a real road profile is uniform and sinusoidal in shape. Any road profile can be decomposed into a series of sinusoidal functions using the Fourier transform. In Fig. 3.12, we can see a typical signal on the left side that can be decomposed into four uniform sinusoidal waves. According to the previous section, dealing with a sinusoidal road profile is easy since we know how to precisely estimate the type of road-induced vibration.

Given the Fourier transform, a real road's profile can be expressed based on a summation of an infinite number of sinusoidal functions as:

$$y(X) = d_0 + \sum_{k=1}^{\infty} d_k \sin(\omega_{pk} X + \varphi_k) \tag{3.7}$$

If we suppose that the tire remains in contact with the road, then Eq. (3.7) can be rewritten as:

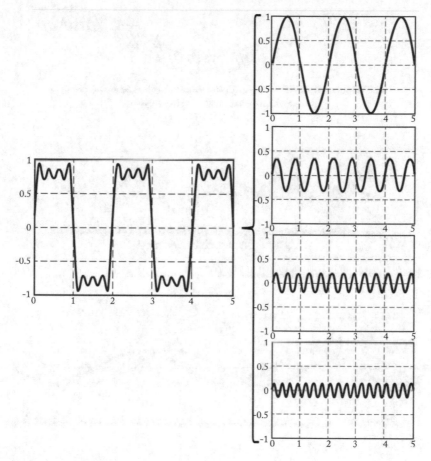

Fig. 3.12 Decomposition of a typical signal to sinc waves

$$y(t) = d_0 + \sum_{k=1}^{\infty} d_k \sin(\omega_k t + \varphi_k) \tag{3.8}$$

where $\omega_k = \frac{2\pi u}{\lambda}$.

c. Standard Road Profile

To assess the ride performance of a car, it is tested first on roads with standard profiles. These roads are discussed below.

Sinusoidal Road Profile: The most relevant kind of road profile that can be seen in test roads is the sinusoidal form. This type of road can be differed according to its spatial frequency. A typical road can have either a low spatial frequency with a long wavelength or a high spatial frequency with a short wavelength. Figure 3.13 shows these two types of standard sinusoidal roads.

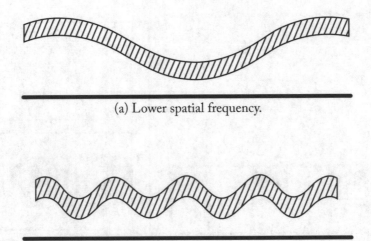

(a) Lower spatial frequency.

(b) Higher spatial frequency.

Fig. 3.13 Two types of standard sinusoidal roads

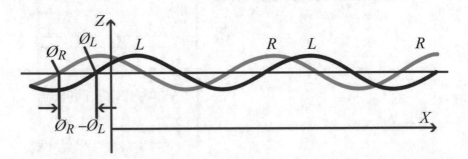

Fig. 3.14 The difference between left and right wheels

The two sides of the test road can be in phase with each other resulting in symmetric motion for each axle wheel. When the two sides of the road are anti-phase, the wheels on an axle move asymmetrically. The different phases between the left and right wheels on one axle are shown in Fig. 3.14.

Thus, the road-induced motion on the left and right wheels can be described by:

$$y_L = d_L \sin\left(\omega_{p_L} X + \varphi_L\right) = d_L \sin(\omega_L t + \varphi_L) \tag{3.9}$$

$$y_R = d_R \sin\left(\omega_{p_R} X + \varphi_R\right) = d_R \sin(\omega_R t + \varphi_R) \tag{3.10}$$

The standard sinusoidal road profiles have different wavelengths, and each wavelength is deliberately designed to comprehensively test a specific behavior of the suspension system. Table 3.1 summarizes different wavelengths and their major influences on a vehicle.

Table 3.1 Major influences on a vehicle with respect to the wide range of a sinusoidal road's wavelength

Characteristic	Wavelength	Influence on
Slopes	100 m $< \lambda$	Static
Undulations	1 m $< \lambda <$ 100 m	Dynamic ride
Roughness	10 mm $< \lambda <$ 1 m	Dynamic NVH
Macro texture	1 mm $< \lambda <$ 10 mm	Friction and noise
Micro texture	10 μm $< \lambda <$ 1 mm	Friction
Material	Molecular	Friction

Isolated Bumps: Another type of standard road profile is isolated bumps, unlike the sinusoidal road profile, which is not continuous. The standard bumps can have different shapes and dimensions; some more important bumps have been named as rectangular, triangular, sine half-wave, and haversine. These standard bumps are shown in Fig. 3.15.

Road excitation according to bump shapes and dimensions can vary greatly, so the motion of the wheel also varies; a wheel's motion upward on a sine half-wave can be formulated as:

$$y - H \sin\left(\frac{\pi X}{L}\right); 0 \leq X \leq L \tag{3.11}$$

While the equation of motion of a wheel on a haversine is defined as (Fig. 3.16):

$$y = H \, hav\left(\frac{2\pi X}{L}\right); 0 \leq X \leq L \tag{3.12}$$

where *hav* function is:

$$hav(\theta) = \frac{1}{2}\{1 - \cos(\theta)\} \tag{3.13}$$

3.4 Ride Comfort Evaluation

Defining a standard technique to evaluate the ride vibration of a vehicle is a debatable topic among automotive engineers. There is still no universal standard for evaluating ride vibrations. However, some standardized methods have been defined to assess the vibration exposure risks considering the human sensitivity to frequency and magnitude of vibration and exposure duration. The most common standard for evaluating ride comfort is ISO 2631–1. This standard evaluates human exposure to whole-body vibration. As an example of vibration exposure assessment recommended by ISO 2631–1, Fig. 3.17 illustrates the decreased proficiency boundaries for vertical and horizontal vibrations in terms of root-mean-square (RMS) of the acceleration as a function of frequency for different

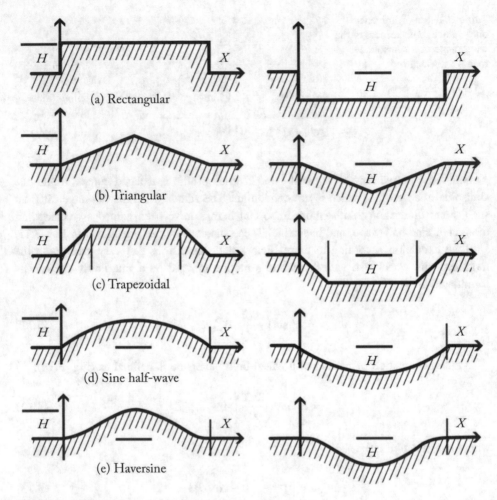

Fig. 3.15 Different shapes of isolated bumps

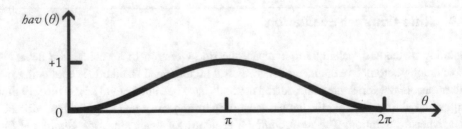

Fig. 3.16 The *hav* function

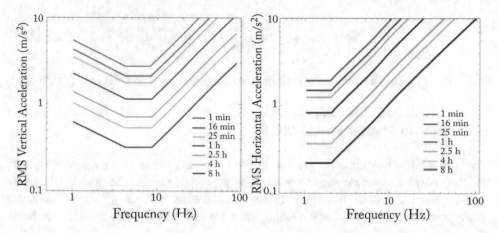

Fig. 3.17 Whole-body vibration limits for decreased proficiency in vertical and horizontal directions, recommended by ISO 2631

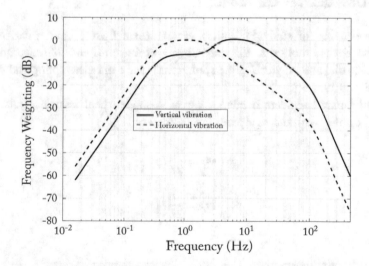

Fig. 3.18 Frequency weighting curves for principal weighting, recommended by ISO 2631

exposure time. As it can be seen, the boundary lowers by increasing the daily exposure time.

The ISO-2631 also defines the frequency weightings for assessing vertical and horizontal vibrations, while the exposure is expressed in terms of frequency-weighted root-mean-square (RMS) acceleration. As shown in Fig. 3.18, the frequency weights used in this standardized exposure assessment method suggest that the human body is most sensitive to horizontal vibration in the 0.5–2 Hz range, and to vertical vibration in the 4–10 Hz range. The weighted RMS acceleration can be formulated as:

$$RMS\left(a_{w(x/y/z)}\right) = \sqrt{\frac{1}{T}\int_0^T \left[a_{w(x/y/z)}\right]^2 dt} \tag{3.14}$$

where $a_{w(x/y/z)}$ denotes frequency-weighted acceleration along a given axis (x or y or z).

3.5 Vehicle Mathematical Model

The spring and shock absorber are two of the main components in a suspension system and they play an important role in vehicle handling and ride quality. Suspension designers usually use the natural frequency method to calculate suitable spring stiffness. To implement the natural frequency method, there are four different ride models introduced below.

3.5.1 1-DOF Quarter Car Model

The simplest model for studying vehicle vibration is a 1 DOF quarter car model. The sprung mass bounce motion is the only degree of freedom, and the unsprung mass is eliminated. A solid mass indicating the sprung mass and a massless spring and damper are shown in Fig. 3.19.

The model only considers a quarter car as a single-wheel vehicle. Thus, the mass distribution on the front and rear axles are defined as:

$$m_f = \frac{b}{2L}M \tag{3.15}$$

$$m_r = \frac{a}{2L}M \tag{3.16}$$

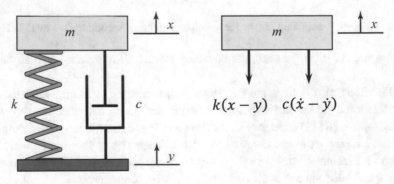

Fig. 3.19 DOF in a quarter car model

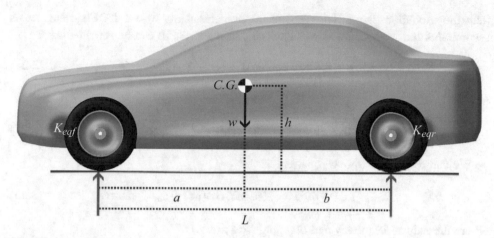

Fig. 3.20 A vehicle's mass distribution over the front and rear axles

where a, b, and L are the dimensions determined in Fig. 3.20.

Moreover, k is the equivalent stiffness of the tire vertical stiffness and suspension spring stiffness in series springs, which is defined as:

$$\frac{1}{k} = \frac{1}{k_s} + \frac{1}{k_t}$$ (3.17)

According to Newton's second law:

$$\sum f = ma$$ (3.18)

Only two forces act on the mass in a simple oscillatory system. One force is due to the spring and the other is due to the shock absorber. Therefore, by assuming $x > y$, the vibration equation of the mass becomes:

$$-c(\dot{x} - \dot{y}) - k(x - y) = m\ddot{x}$$ (3.19)

or

$$m\ddot{x} + c(\dot{x} - \dot{y}) + k(x - y) = 0$$ (3.20)

Transmissibility refers to the ratio of the maximum amplitude of the induced signal transmitted to the mass to the amplitude of the excitation signal. If transmissibility equals one, the mass will vibrate at the same amplitude as the excitation signal. If it is less than 1, the amplitude of the mass will be smaller than the excitation signal. In addition, if it is higher than 1, the system amplifies the mass vibration, and it will have a larger amplitude than the excitation signal. To calculate the transmissibility of a 1 DOF system, let us assume that the road profile is $y = Y\sin(\omega t)$, then Eq. (3.20) can be rewritten as:

$$m\ddot{x} + c(\dot{x} - Y\omega\cos(\omega t)) + k(x - Y\sin\omega t) = 0 \qquad (3.21)$$

or

$$m\ddot{x} + c\dot{x} + kx = c\omega Y\cos(\omega t) + kY\sin\omega t \qquad (3.22)$$

We can rewrite the above equation as:

$$\ddot{x} + 2\zeta\omega_n\dot{x} + \omega_n^2 x = 2\zeta\omega_n\omega Y\cos(\omega t) + \omega_n^2 Y\sin\omega t \qquad (3.23)$$

where the natural frequency and damping ratio are:

$$\omega_n = \sqrt{\frac{k}{m}} = 2\pi f_n \qquad (3.24)$$

$$\zeta = \frac{c}{2\sqrt{km}} \qquad (3.25)$$

If the sprung mass oscillates as:

$$x = X\sin(\omega t - \varphi_x) \qquad (3.26)$$

then, the steady state amplitude transmissibility can be obtained as:

$$\left|\frac{X}{Y}\right| = \sqrt{\frac{1 + (2\zeta r)^2}{(1 - r^2)^2 + (2\zeta r)^2}} \qquad (3.27)$$

in which

$$r = \frac{\omega}{\omega_n} \qquad (3.28)$$

The response of the steady-state amplitude transmissibility with respect to the proportion of the induced frequency and natural frequency is shown in Fig. 3.21.

As shown in Fig. 3.21, if the damping ratio ζ is zero and the frequency of excitation is equal to the natural frequency of the system ($\omega/\omega_n = 1$) the amplitude will increase to infinity. This is called resonance. Moreover, for any damping ratio, the transmissibility is 1 at the frequency of $\sqrt{2}\omega_n$, which means that the mass amplitude is equal to the road profile amplitude. For frequencies higher than this, the transmissibility becomes less than 1.

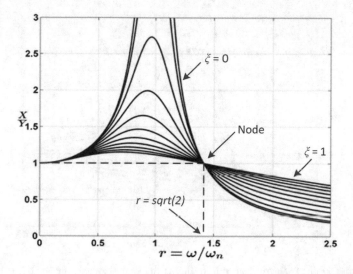

Fig. 3.21 Displacement transmissibility of the sprung mass

In addition, in case we want to study the relative amplitude between the mass and road, we can consider:

$$z = Z \sin(\omega t - \varphi_z), \; z = x - y \tag{3.29}$$

Therefore, Eq. (3.23) can be written as:

$$\ddot{z} + 2\zeta\omega_n\dot{z} + \omega_n^2 z = \omega^2 Y \sin(\omega t) \tag{3.30}$$

Finally, the steady-state relative amplitude transmissibility can be obtained as:

$$\left|\frac{Z}{Y}\right| = \frac{r^2}{\sqrt{(1 - r^2)^2 + (2\zeta r)^2}} \tag{3.31}$$

The mass relative amplitude transmissibility with respect to the normalized excitation frequency (ω/ω_n) is plotted in Fig. 3.22.

As shown in Fig. 3.22, if the damping ratio ζ is zero, the relative amplitude will increase to infinity when the excitation frequency is equal to the system's natural frequency $(\omega/\omega_n = 1)$. As the excitation frequency increases, the transmissibility approaches 1, which means that the mass amplitude equals to the road profile amplitude. The transmissibility of systems with lower damping ratios is larger than systems with larger damping ratios.

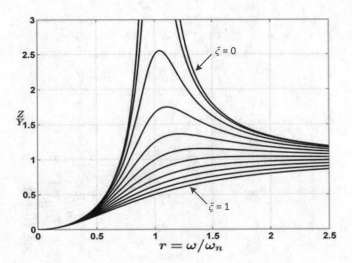

Fig. 3.22 Relative displacement transmissibility of the sprung mass

If we want to study sprung mass acceleration, we can define the acceleration transmissibility as:

$$\left|\frac{\ddot{X}}{Y\omega_n{}^2}\right| = \frac{r^2\sqrt{1+(2\zeta r)^2}}{\sqrt{(1-r^2)^2+(2\zeta r)^2}} \tag{3.32}$$

The acceleration transmissibility with respect to the normalized excitation frequency (ω/ω_n) is shown in Fig. 3.23.

According to Fig. 3.23, the transmissibility increases to infinity when there is no damping in the system ($\zeta = 0$) at the resonance frequency ($\omega/\omega_n = 1$). In addition, when $\omega/\omega_n = \sqrt{2}$, the transmissibility equals 2 and is independent from the damping ratio as shown in the figure. For excitation frequencies higher than $\omega = \sqrt{2}\omega_n$, the lower the damping ratio, the lower the acceleration of the mass. Regardless of the damping ratio, however, the acceleration transmissibility is larger than 1 for frequencies above $\sqrt{2}\omega_n$.

The phase and relative phase of the sprung mass can be defined as:

$$\varphi_x = arc\tan\left(\frac{2\zeta r^3}{1-r^2+(2\zeta r)^2}\right) \tag{3.33}$$

$$\varphi_z = arc\tan\left(\frac{2\zeta r}{1-r^2}\right) \tag{3.34}$$

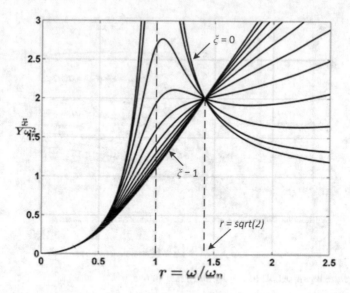

Fig. 3.23 Acceleration transmissibility of the sprung mass

where φ_x and φ_z are the relative phase of sprung mass motion with respect to the excitation phase. The frequency response of the sprung mass phase is shown in Fig. 3.24.

Example 3.1: Consider the quarter car model in Fig. 3.19 with the following specifications (Table 3.2):

(a) Find the natural frequency and damping ratio of the suspension.
(b) Plot the amplitude transmissibility, relative amplitude transmissibility, acceleration transmissibility, and the phase of the sprung mass.
(c) Simulate and plot the time response (displacement and acceleration) of the 1 DOF quarter car model to a typical haversine bump for a vehicle speed of 10 km/h. Discuss how the mass can affect the results (Fig. 3.25).

Solution

(a) To find the natural frequency and damping ratio of the suspension using a quarter car model or a 1 DOF model, the equivalent stiffness of the tire and spring should be calculated from Eq. (3.17):

$$\frac{1}{k} = \frac{1}{28} + \frac{1}{240}$$

$$k = 25.075 \text{ kN/m}$$

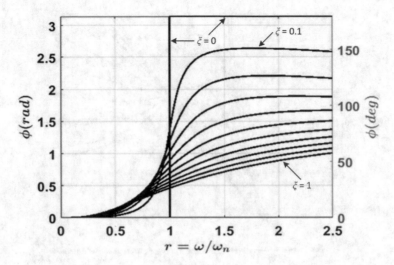

Fig. 3.24 Frequency response of the sprung mass phase

Table 3.2 Specifications of a quarter car model

Symbol	Parameter	Value
m_s	Sprung mass	400 kg
m_u	Unsprung mass	45 kg
k_s	Spring stiffness	28 kN/m
c	Damping coefficient	2500 Ns/m
k_t	Tire stiffness	240 kN/m

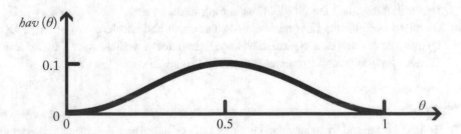

Fig. 3.25 A typical haversine

The natural frequency and damping ratio can be obtained using Eqs. (3.24) and (3.25), respectively:

$$\omega_n = \sqrt{\frac{k}{m_s + m_u}} = \sqrt{\frac{25075}{445}} = 7.51 \frac{rad}{s} = 1.19 \text{ Hz}$$

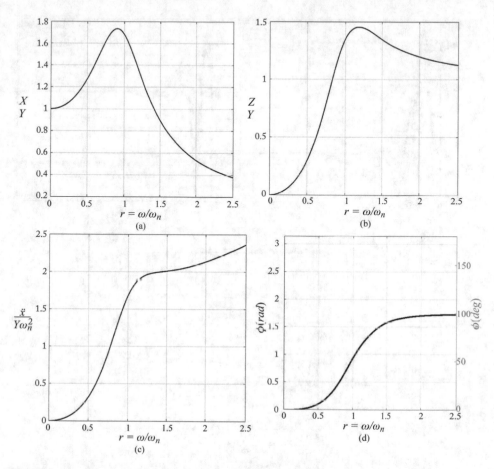

Fig. 3.26 Frequency response of a 1 DOF quarter model: **a** amplitude transmissibility **b** relative amplitude transmissibility **c** acceleration transmissibility **d** sprung mass's phase

$$\zeta = \frac{2500}{2\sqrt{25075 \times 445}} = 0.3742$$

(b) The frequency response of the 1 DOF quarter model can be plotted using Eqs. (3.27), (3.31), (3.32), and (3.33), as shown in Fig. 3.26.

(c) The time response of the 1 DOF quarter model can be determined by solving the differential equation of this model presented in Eq. (3.20). Figure 3.27 illustrates the time response of the model for three different values of mass. It can be seen that whenever the mass is lighter, the amplitude of vibration decreases more quickly because the lower masses require less kinematic energy so the shock absorber can turn this energy into heat more quickly.

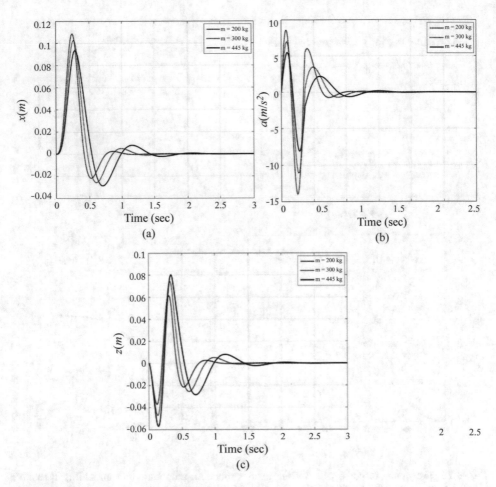

Fig. 3.27 Time response of a 1 DOF quarter model to a typical haversine bump: **a** displacement **b** acceleration **c** relative displacement

Example 3.2: For the 1 DOF quarter model described in Example 2.1, consider two dampers with a higher damping coefficient of $c = 4010$ Ns/m and a lower damping coefficient of $c = 1005$ Ns/m.

(a) Find the frequency response of the model and compare the results with those obtained in Example 2.1.
(b) After plotting the time response of the 1 DOF quarter model to a sine function with an amplitude of 0.1 m and frequencies of 0.5 and 5 Hz, discuss how choosing a higher or lower damping coefficient can affect model response in low and high frequencies.

Solution

(a) Using Eq. (3.25) the damping ratio for $c = 4010$ Ns/m and $c = 1005$ Ns/m are 0.60 and 0.15, respectively.

The amplitude and acceleration transmissibility of the sprung mass for the above damping ratios and Example 3.1 are shown in Fig. 3.28. As can be seen in the Figure, increasing the damping ratio improves the ride quality at low frequencies. This means that both displacement transmissibility and acceleration transmissibility are lower at frequencies below $r = \sqrt{2}$. On the other hand, a suspension system with a lower damping ratio performs better at high frequencies (i.e., $r > \sqrt{2}$). As a result, choosing an optimum damping ratio for the suspension system is critical. This parameter should be determined such that the suspension system has an acceptable performance in both low and high frequencies.

(b) As discussed in part (a), a higher damping ratio is more effective in low frequencies. The time response of the model to a sine wave with a low frequency of 0.5 Hz confirms that the suspension system with $\zeta = 0.6$ has the smallest displacement and acceleration. However, for a sine wave with a high frequency of 5 Hz, the lowest damping ratio ($\zeta = 0.15$) has the smallest displacement and acceleration. The time response of the suspension system with $\zeta = 0.37$ shows good performance in both cases. Therefore, $\zeta = 0.37$ is a better option for the damping ratio of this suspension system (Fig. 3.29).

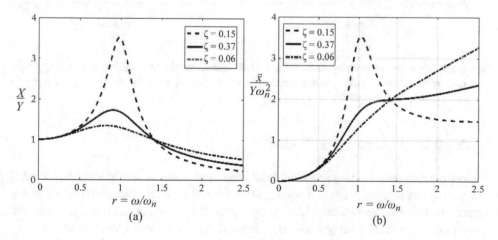

Fig. 3.28 Frequency response comparison of a 1 DOF quarter model with different damping ratios: **a** amplitude transmissibility **b** acceleration transmissibility

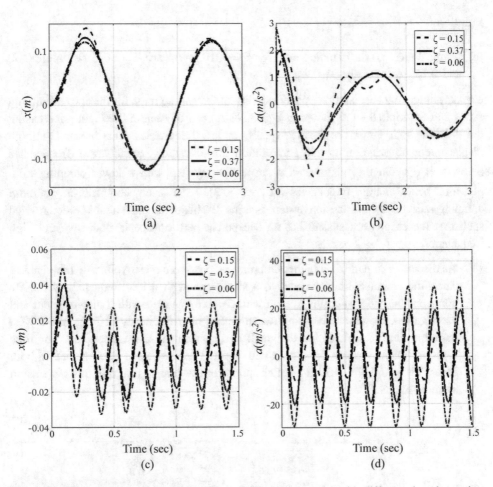

Fig. 3.29 Time response comparison of a 1 DOF quarter model with different damping ratios: **a** displacement at low frequency **b** acceleration at low frequency **c** displacement at high frequency **d** acceleration at high frequency

3.5.2 2-DOF Quarter Car Model

A more complicated quarter-car model has 2 degrees of freedom. As shown in Fig. 3.30, it includes both unsprung mass and sprung mass bounce motions. The unsprung mass m_{us} represents the wheel and its associated components, and the sprung mass m_s represents the corresponding vehicle's body mass to the wheel (approximately a quarter of the body mass). Sprung and unsprung bounce motions are described by x_1 and x_2, respectively. Suspension stiffness and damping coefficient are k_s and c_s. Tire vertical stiffness is k_t and its damping is usually negligible with respect to the suspension damping.

Fig. 3.30 2-DOF quarter car model

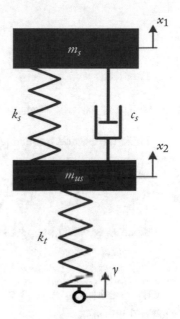

Based on Fig. 3.30 and assuming $x_2(t) > x_1(t)$, the equations of motion can be obtained using Newton's second law as:

$$m_s \ddot{x}_1(t) = c_s[\dot{x}_2(t) - \dot{x}_1(t)] + k_s[x_2(t) - x_1(t)] \tag{3.35}$$

$$m_{us} \ddot{x}_2(t) = -c_s[\dot{x}_2(t) - \dot{x}_1(t)] - k_s[x_2(t) - x_1(t)] - k_t[x_2(t) - y(t)] \tag{3.36}$$

To find the frequency response of the system, we assume:

$$x_1(t) = X_s \sin(\omega t - \varphi_1) \tag{3.37}$$

$$x_2(t) = X_{us} \sin(\omega t - \varphi_2)$$
$$y(t) = Y \sin \omega t \tag{3.38}$$

By replacing Eqs. (3.37) and (3.38) in Eqs. (3.35) and (3.36), we obtain:

$$(-m_s \omega^2 + c_s \omega + k_s)X_s + (-c_s \omega - k_s)X_{us} = 0 \tag{3.39}$$

$$(-c_s \omega - k_s)X_s + (-m_{us} \omega^2 + c_s \omega + k_s + k_t)X_{us} = k_t Y \tag{3.40}$$

To determine the undamped natural frequency of the system, it is assumed that $c_s = 0$ and $Y = 0$. To have a non-trivial solution for the system of Eqs. (3.39) and (3.40), we need to have:

$$det \begin{bmatrix} -m_s\omega^2 + k_s & -k_s \\ -k_s & -m_{us}\omega^2 + k_s + k_t \end{bmatrix} = 0 \qquad (3.41)$$

This leads to:

$$m_s m_{us}\omega^4 - [(k_s + k_t)m_s + k_s m_{us}]\omega^2 + k_s k_t = 0 \qquad (3.42)$$

Thus, the natural frequency can be determined as:

$$\omega^2 = \frac{k_s + k_t}{2m_{us}} + \frac{k_s}{2m_s} \pm \frac{\sqrt{(k_s + k_t)^2 m_s^2 + m_{us}^2 k_s^2 - 2(k_t - k_s)k_s m_s m_{us}}}{2m_s m_{us}} \qquad (3.43)$$

Since k_t is vastly bigger than k_s, it is assumed that:

$$k_s + k_t \approx k_t - k_s \approx k_t \qquad (3.44)$$

By considering Eqs. (3.43) and (3.44), the approximate natural frequencies of the system become:

$$\omega_{n_s} = \sqrt{\frac{k_s}{m_s}} \qquad (3.45)$$

$$\omega_{n_{us}} = \sqrt{\frac{k_t}{m_{us}}} \qquad (3.46)$$

Usually, for a modern passenger car, the natural frequency of the sprung mass is approximately 1 Hz, and the natural frequency of the unsprung mass is approximately 10 Hz.

Assuming m_{s_v} as the vehicle sprung mass, the front and rear axle mass can be determined by:

$$m_{s_f} = \frac{m_{s_v} b}{L} \qquad (3.47)$$

$$m_{s_r} = \frac{m_{s_v} a}{L} \qquad (3.48)$$

where m_{s_f} and m_{s_r} are the mass distribution over the front and rear axle. By using Eqs. (3.45), (3.47), and (3.48), the spring stiffness in the front and rear axles will become:

$$k_{S_{front}} = \frac{m_{s_v} b}{2L}(\omega_{n_s})^2 \qquad (3.49)$$

$$k_{S_{rear}} = \frac{m_{s_v} a}{2L} (\omega_{n_s})^2 \tag{3.50}$$

To improve the vehicle pitch motion, the sprung mass natural frequency in the rear axle is bigger than in the front axle, therefore, the natural frequency of the front axle is chosen to be 1 Hz and the natural frequency of the rear axle is selected to be 1.2 Hz.

In order to determine amplitude transmissibility, Eqs. (3.39) and (3.40) are used and the sprung mass amplitude transmissibility becomes:

$$\mu = \left| \frac{X_s}{Y} \right|$$

$$\mu^2 = \frac{4\xi^2 r^2 + 1}{Z_1^2 + Z_2^2} \tag{3.51}$$

in which

$$Z_1 = \left[r^2 (r^2 \alpha^2 - 1) + (1 - (1+\varepsilon) r^2 \alpha^2) \right] \tag{3.52}$$

$$Z_2 = 2\zeta r (1 - (1+\varepsilon) r^2 \alpha^2) \tag{3.53}$$

$$\xi = \frac{c_s}{2 m_s \omega_s} \tag{3.54}$$

$$\alpha = \frac{\omega_s}{\omega_u} \tag{3.55}$$

$$\varepsilon = \frac{m_s}{m_u} \tag{3.56}$$

$$r = \frac{\omega}{\omega_n} \tag{3.57}$$

The response of the sprung mass amplitude transmissibility with respect to the proportion of the induced frequency and the natural frequency is shown in Fig. 3.31.

The system has 2 DOF and as Fig. 3.31 shows, it has two natural frequencies where the amplitude of the system without a damping ratio will go to infinity. The natural frequencies, r_{n_1} and r_{n_2}, can be determined as:

$$r_{n_1} = \sqrt{\frac{1}{2\alpha^2} \left(1 + (1+\varepsilon)\alpha^2 - \sqrt{(1 + (1+\varepsilon)\alpha^2)^2 - 4\alpha^2} \right)} \tag{3.58}$$

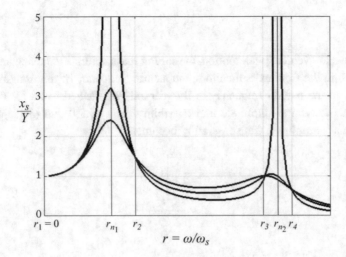

Fig. 3.31 The sprung mass amplitude transmissibility

$$r_{n_2} = \sqrt{\frac{1}{2\alpha^2}\left(1 + (1+\varepsilon)\alpha^2 + \sqrt{(1+(1+\varepsilon)\alpha^2)^2 - 4\alpha^2}\right)} \qquad (3.59)$$

As shown in Fig. 3.31, in r_2, r_3, and r_4, the transmissibility is independent of the damping values. Furthermore, between r_2 and r_3, a sprung mass with a lower damping ratio has smaller transmissibility. However, it has bigger transmissibility in the range of r_1 and r_2.

The unsprung mass amplitude transmissibility is:

$$\tau = \left|\frac{X_u}{Y}\right|$$
$$\tau^2 = \frac{4\xi^2 r^2 + 1 + r^2(r^2 - 2)}{Z_1^2 + Z_2^2} \qquad (3.60)$$

In addition, the relative amplitude transmissibility between the sprung and unsprung masses can be defined as:

$$\eta = \left|\frac{Z}{Y}\right|$$
$$\eta^2 = \frac{r^4}{Z_1^2 + Z_2^2} \qquad (3.61)$$

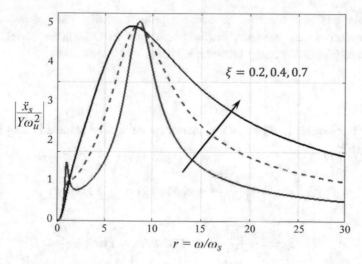

Fig. 3.32 The response of the steady state sprung acceleration transmissibility

Moreover, the acceleration transmissibility for the sprung and unsprung masses can be defined as:

$$\left| \frac{\ddot{X}_s}{Y\omega_s^2} \right| = r^2 \alpha^2 \mu \tag{3.62}$$

$$\left| \frac{\ddot{X}_u}{Y\omega_u^2} \right| = r^2 \alpha^2 \tau \tag{3.63}$$

The response of the sprung mass acceleration transmissibility with respect to the proportion of the induced frequency and natural frequency is shown in Fig. 3.32. The system has two DOFs, so it has two natural frequencies, and the two jumps are shown in Fig. 3.32, which also indicates how systems with lower damping ratios have more critical situations in these frequencies even though they have smaller transmissibility in other frequencies. One of the natural frequencies is approximately 1, which refers to the sprung mass, and the other is approximately 10, which is associated with the unsprung mass.

Example 3.3: Using the suspension specifications provided in Example 3.1 and a two DOFs quarter model:

(a) Find the natural frequencies and the damping ratio of the suspension.
(b) Plot the frequency response (amplitude transmissibility and acceleration transmissibility) of the sprung mass and unsprung mass along with the relative amplitude transmissibility.

(c) Simulate and plot the time response (sprung/unsprung mass displacement and acceleration as well as relative displacement) of a 2 DOF quarter model to a typical haversine (Fig. 3.30) bump for vehicle speeds of 10 km/h.

Solution

(a) Natural frequencies in a two DOFs quarter model can be obtained using Eqs. (3.45) and (3.46):

$$\omega_s = \sqrt{\frac{28000}{400}} = 8.36 \text{ rad/s} = 1.33 \text{ Hz}$$

$$\omega_u = \sqrt{\frac{240000}{45}} = 73.03 \text{ rad/s} = 11.62 \text{ Hz}$$

Using Eqs. (3.55) to (3.59), the natural frequencies r_{n_1} and r_{n_2} can be determined as:

$$\alpha = \frac{\omega_s}{\omega_u} = \frac{8.36}{73.03} = 0.11$$

$$\varepsilon = \frac{m_s}{m_u} = 8.88$$

$$r_{n1} = \sqrt{\frac{1}{2(0.11)^2}\left(1+(1+8.88)(0.11)^2\right) - \sqrt{\left(1+(1+8.88)(0.11)^2\right)^2 - 4(0.11)^2}}$$
$$= 0.95 \text{ Hz}$$

$$r_{n1} = \sqrt{\frac{1}{2(0.11)^2}\left(1+(1+8.88)(0.11)^2\right) + \sqrt{\left(1+(1+8.88)(0.11)^2\right)^2 - 4(0.11)^2}}$$
$$= 9.23 \text{ Hz}$$

The damping ratio in this model can be calculated by Eq. (3.54):

$$\xi = \frac{c_s}{2m_s\omega_s} = \frac{2500}{2 \times 400 \times \sqrt{\frac{28000}{400}}} = 0.3735$$

(b) As shown in Fig. 3.33, the frequency response of the 2 DOF quarter model can be plotted using Eqs. (3.51), (3.60) to (3.63). As expected, the highest amplitude of the sprung mass occurs at the first natural frequency, and the highest amplitude of the unsprung mass happens at the second natural frequency.

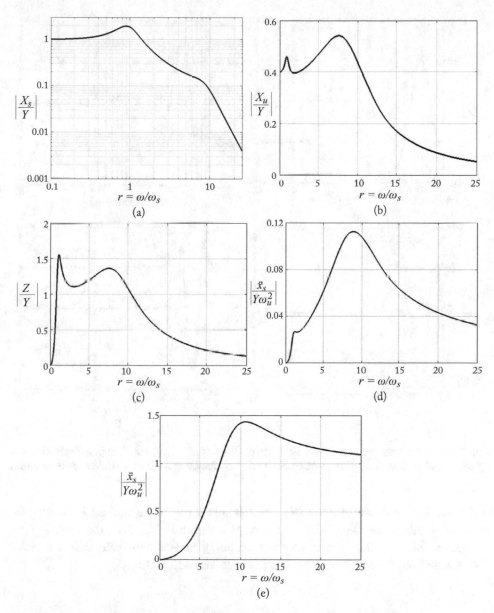

Fig. 3.33 Frequency response of a 2-DOF quarter car model: **a** sprung mass amplitude transmissibility **b** unsprung mass amplitude transmissibility **c** relative amplitude transmissibility **d** sprung mass acceleration transmissibility **e** unsprung mass acceleration transmissibility

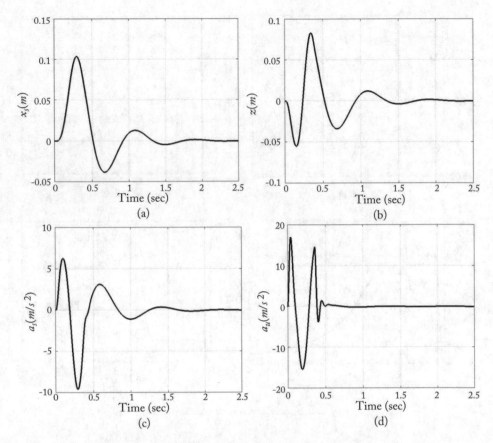

Fig. 3.34 Time response of a 2-DOF quarter model to a typical haversine bump: **a** displacement of sprung mass, **b** relative displacement, **c** acceleration of sprung mass, **d** acceleration of the unsprung mass

(c) The time response of the 2 DOF quarter model can be determined by solving the differential equations of this model. As it is shown in Fig. 3.34, the haversine isolated bump induces the sprung and unsprung masses to vibrate, and after some oscillation, the shock absorber dampens the kinematic energy.

3.5.3   Half-car Model

A half-car model includes the bounce and pitch motions of the body as shown in Fig. 3.35. To determine the undamped natural frequencies, the suspension and tire damping are neglected and the equivalent suspension and tire stiffness, k_{eq} is defined as:

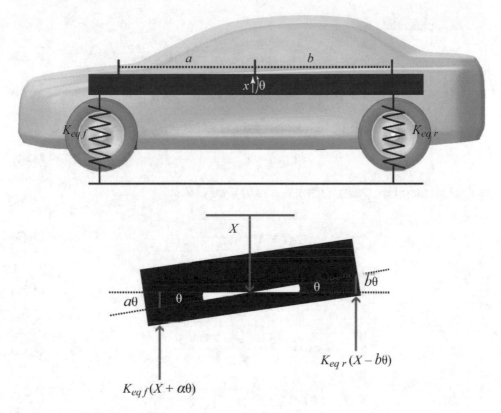

Fig. 3.35 Half-car ride model

$$k_{eq} = \frac{k_s k_t}{k_s + k_i} \tag{3.64}$$

The acting forces on the body of the vehicle are shown in Fig. 3.35. According to Newton's second law, the equations of motion are written as:

$$\sum F_x = ma_x \tag{3.65}$$

$$\sum M_{C.G} = I_{yy}\ddot{\theta} \tag{3.66}$$

where m and I_{yy} are half of the vehicle's mass and pitch mass inertia, respectively. Equations (3.65) and (3.66) can be rewritten as:

$$-k_{eq\,f}(x + a\theta) - k_{eq\,r}(x - b\theta) = m\ddot{x} \tag{3.67}$$

$$-k_{eq\,f}(x + a\theta)a + k_{eq\,r}(x - b\theta)b = I_{C.G}\ddot{\theta} \tag{3.68}$$

By simplification:

$$m\ddot{x} + \left(k_{eqf} + k_{eq\,r}\right)x + \left(k_{eqf}a - k_{eq\,r}b\right)\theta = 0 \tag{3.69}$$

$$I_{C.G}\ddot{\theta} + \left(k_{eqf}a^2 + k_{eq\,r}b^2\right)\theta + \left(k_{eqf}a - k_{eq\,r}b\right)x = 0 \tag{3.70}$$

Using the moment equilibrium around the CG of the vehicle in steady states:

$$k_{eqf}a - k_{eq\,r}b = 0 \tag{3.71}$$

and considering Eqs. (3.69), (3.70), and (3.71), we obtain:

$$\omega_{n,T} = \sqrt{\frac{k_{eqf} + k_{eq\,r}}{m}} \tag{3.72}$$

$$\omega_{n,R} = \sqrt{\frac{k_{eqf}a^2 + k_{eq\,r}b^2}{I_{C.G}}} \tag{3.73}$$

$$k_{eqf} = \frac{I_{C.G}\omega_{n,R}^2 - b^2 m\omega_{n,T}^2}{a^2 - b^2} \tag{3.74}$$

$$k_{eq\,r} = m\omega_{n,T}^2 - \frac{I_{C.G}\omega_{n,R}^2 - b^2 m\omega_{n,T}^2}{a^2 - b^2} \tag{3.75}$$

Usually, the bounce natural frequency $\omega_{n,T}$ and pitch natural frequency $\omega_{n,R}$ have a suggested value of approximately 1 Hz.

Example 3.4: Consider an electric vehicle with two in-wheel motors. The car's batteries are accommodated in the vehicle floor structure. The vehicle mass is 920 kg and its pitch inertia I_{yy} is 745 kgm^2, while each wheel and its associated mass is 12 kg, and each electric motor mass is 20 kg. The vehicle wheelbase is 1.8 m and the distance between the center of gravity and rear axle is $b = 0.895$ m. The tire vertical stiffness is $k_t = 220$ kN/m.

(a) Find the front and rear spring stiffness using a quarter car model.
(b) Find the front and rear spring stiffness using a half-car model.

Solution

(a) The vehicle wheelbase is 1.8 m and $b = 0.895$ m, so $a = 0.905$ m.

Electric motors are installed in the rear wheel and are counted in the unsprung mass, so in the quarter car model:

$$m_{us_f} = 12 \text{ kg}$$

$$m_{us_r} = 32 \text{ kg}$$

and

$$m_{s_v} = m - m_{us} = 920 - 2 \times 12 - 2 \times 32 = 832 \text{ kg}$$

Usually, the spring stiffness in the rear axle is bigger than in the front axle. So, the natural frequency in the front axle is assumed to be 1 Hz, and in the rear axle, it is selected as 1.2 Hz.

Note that 1 Hz $= 2\pi$ rad/sec, by using Eqs. (3.49) and (3.50):

$$k_{S_{front}} = \frac{m_{s_v}b}{2L}(\omega_{n_s})^2 = \frac{832 \times 0.895}{2 \times 1.8}(2\pi)^2 \approx 8,166 \text{ N/m}$$

$$k_{S_{rear}} = \frac{m_{s_v}a}{2L}(\omega_{n_s})^2 = \frac{832 \times 0.905}{2 \times 1.8}(1.2 \times 2\pi)^2 \approx 11,890 \text{ N/m}$$

(b) The bounce and pitch natural frequencies are selected as 1 Hz. So, by using Eqs. (3.74) and (3.75):

$$k_{eqf} = \frac{I_{CG}\omega_{n,R}^2 - b^2 m\omega_{n,T}^2}{a^2 - b^2} = \frac{745.18 \times (2\pi)^2 - 0.895^2 \times 920 \times (2\pi)^2}{2(0.905^2 - 0.895^2)} \approx 9033 \text{ N/m}$$

$$k_{eq\,r} = m\omega_{n,T}^2 - \frac{I_{CG}\omega_{n,R}^2 - b^2 m\omega_{n,T}^2}{2(a^2 - b^2)} = m\omega_{n,T}^2 - k_{eqf} = 920 \times (2\pi)^2 - k_{eqf} \approx 9127 \text{ N/m}$$

According to Eq. (3.64):

$$k_{S_f} = \frac{k_t k_{eqf}}{k_t - k_{eqf}} = \frac{220000 \times 9033}{220000 - 9033} \approx 9420 \text{ N/m}$$

$$k_{S_r} = \frac{k_t k_{eq\,r}}{k_t - k_{eq\,r}} = \frac{220000 \times 9127}{220000 - 9127} \approx 9522 \text{ N/m}$$

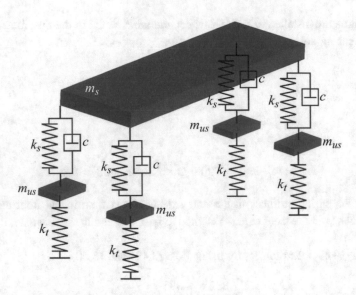

Fig. 3.36 Full car ride model

3.5.4   Full-car Model

A full car ride model has 7 degrees of freedom including bounce, pitch, and roll motions for the vehicle body, and 4 bounce motions for the four wheels. This model is too complicated for natural frequency analysis, but it is applicable for accurate computer simulations. Figure 3.36 shows a full car model diagram.

3.6   Installation Effects

In the above models, the spring is supposed to be installed directly above the tire. However, in reality, the spring is connected to the control arm that needs to be considered in the calculation of the effective spring and damping coefficient. Using Fig. 3.37, we can write:

$$\frac{\Delta z}{\Delta z_a} = \frac{l}{l_1} \tag{3.76}$$

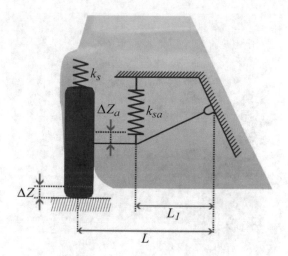

Fig. 3.37 Spring installation effects

then

$$\Delta z_a = \frac{l_1}{l}\Delta z \qquad (3.77)$$

The potential energy that is stored in the theoretical and real springs should equal:

$$E = E_a \qquad (3.78)$$

$$\frac{1}{2}k_s\Delta z^2 = \frac{1}{2}k_{sa}\Delta z_a^2 \qquad (3.79)$$

By substituting Eqs. (3.77) in (3.79):

$$k_{sa} = k_s\left(\frac{l}{l_1}\right)^2 \qquad (3.80)$$

Example 3.5: For the front suspension of the vehicle in Example 3.4a, find the actual spring stiffness based on the following figure:

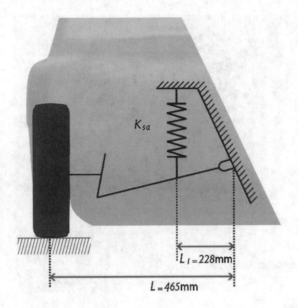

Solution

According to Example 3.4a, the calculated front spring stiffness is 8166 N/m. By using Eq. (3.80), the actual spring stiffness is:

$$k_{sa} = k_s \left(\frac{l}{l_1}\right)^2 = 8166 \left(\frac{0.465}{0.288}\right)^2 \approx 21287 \text{ N/m}$$

3.7   Anti-roll Bar Sizing

The main function of the anti-roll bar is to control the vehicle's roll during cornering maneuvers. To study the vehicle's roll motion, a one degree of freedom vehicle roll model is shown in Fig. 3.38. The body rolls about the roll center R_C while the lateral inertia force acts on the CG during cornering. The roll model can be represented as an inverted pendulum. The spring and the shock absorber are modeled as a torsional spring k_{ta} and a torsional damper c_{ta}, where they respectively, represent torsional stiffness and torsional damping of the vehicle.

By using the D'Alembert principle, the momentum equation about the CG is:

$$-m_s a_y h \cos \varphi - m_s g h \sin \varphi + k_t \varphi + c_t \dot{\varphi} = -I_{zs} \ddot{\varphi} \tag{3.81}$$

where I_{zs} is the roll momentum of inertia of the sprung mass, and it is defined as:

$$I_{zs} = I + mh^2 \tag{3.82}$$

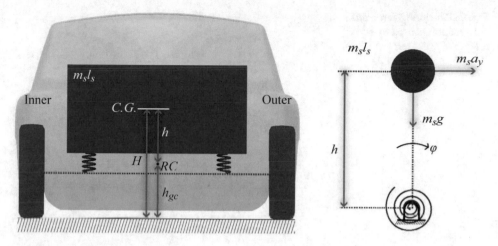

Fig. 3.38 Vehicle roll model

where I is the roll momentum of inertia of the sprung mass around the CG. For a small roll angle φ, we can assume:

$$\sin \varphi \approx \varphi$$
$$\cos \varphi \approx 1 \tag{3.83}$$

As such, Eq. (3.81) is rewritten as:

$$I_{zs}\ddot{\varphi} + c_t\dot{\varphi} + (k_t - m_sgh)\varphi = m_sa_yh \tag{3.84}$$

The roll angle φ is determined numerically by solving Eq. (3.84). In the steady state condition, we can suppose:

$$\ddot{\varphi} = \dot{\varphi} = 0 \tag{3.85}$$

As such, Eq. (3.84) is simplified to:

$$(k_t - m_sgh)\varphi_{ss} = m_sa_yh \tag{3.86}$$

where φ_{ss} is the steady state roll angle. By rewriting Eq. (3.86):

$$\frac{\varphi_{ss}}{a_y} = \frac{m_sh}{k_t - m_sgh} \tag{3.87}$$

Fig. 3.39 Spring force and deformation caused by body roll

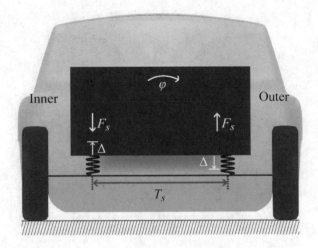

where φ_{ss}/a_y is the roll gain with a normal range of 4 to 8 deg/g. By using Eq. (3.87), the vehicle torsional stiffness is found as:

$$k_t = \frac{m_s h}{\varphi_{ss}/a_y} + m_s g h \tag{3.88}$$

To formulate the vehicle torsional stiffness, consider a vehicle during cornering: the vehicle body rolls about the roll axis and the outer spring compresses as the inner spring expands. The spring force direction and the spring deformation δ are shown in Fig. 3.39.

For a small roll angle, the spring deformation is determined as:

$$\Delta \approx \frac{\varphi T_s}{2} \tag{3.89}$$

The generated roll momentum by the springs of an axle is:

$$M_t = F_s T_s \tag{3.90}$$

where the spring force F_s is:

$$F_s = k_s \Delta \tag{3.91}$$

By substituting Eqs. (3.89) in (3.91):

$$F_s = k_s \frac{\varphi T_s}{2} \tag{3.92}$$

and by substituting Eqs. (3.92) in (3.90):

$$M_t = k_{ta}\varphi \tag{3.93}$$

in which

$$k_{ta} = \frac{1}{2}k_s T_s^2 \tag{3.94}$$

the total torsional stiffness of the vehicle is calculated as:

$$(k_t)_{vehicle} = \left(\frac{1}{2}k_s T_s^2\right)_f + \left(\frac{1}{2}k_s T_s^2\right)_r + (k_t)_{ANR_f} + (k_t)_{ANR_r} \tag{3.95}$$

The total torsional stiffness of the vehicle is the summation of the spring torsional stiffness and the anti-roll bar torsional stiffness for the front and rear axles. Using an anti-roll bar in the rear axle is not common; therefore, by assuming the rear anti-roll bar stiffness is zero, the front anti-roll-bar stiffness is the only unknown variable in Eq. (3.95):

$$(k_t)_{ANR_f} = (k_t)_{vehicle} - \left(\frac{1}{2}k_s T_s^2\right)_f - \left(\frac{1}{2}k_s T_s^2\right)_r \tag{3.96}$$

Example 3.6: Determine the front anti-roll bar torsional stiffness of a vehicle with the following data:

- Vehicle mass $m = 1340$ kg,
- Vehicle sprung mass $m_s = 1287$ kg,
- Height of vehicle center of gravity $H = 0.506$ m,
- Height of vehicle roll axis $h_{RC} = 0.335$ m,
- Distance between the front axle and the center of gravity $a = 1.135$ m,
- Distance between the rear axle and the center of gravity $b = 1.415$ m,
- Front and rear spring stiffness $k_{sf} = 20$ kN/m and $k_{sr} = 22$ kN/m,
- Distance between the springs on the front and rear axles $T_{sf} = 1.115$ m and $T_{sr} = 0.925$ m.

Solution

The distance between the vehicle's center of gravity and the vehicle roll axis is:

$$h = H - h_{RC} = 0.171 \text{ m}$$

The roll gain is supposed to be 6 deg/g, so:

$$\frac{\varphi}{a_y} = 6 \ deg/g \times \frac{\pi}{180} \ rad/deg \times \frac{1}{9.81} \ g/m/s^2 = 0.01 \ rads^2/m$$

By using Eq. (3.88), the vehicle torsional stiffness is calculated as:

$$k_t = \frac{m_s h}{\varphi_{ss}/a_y} + m_s gh = \frac{1287 \times 0.171}{0.01} + 1287 \times 9.81 \times 0.171 = 24167 \ Nm$$

The front and rear torsional stiffness are:

$$(k_t)_{front} = \frac{1}{2} k_{sf} T_{sf}^2 = \frac{1}{2} \times 20000 \times 1.115^2 = 12432 \ Nm$$

$$(k_t)_{rear} = \frac{1}{2} k_{sr} T_{sr}^2 = \frac{1}{2} \times 22000 \times 0.925^2 = 9412 \ Nm$$

Thus, according to Eq. (3.96), the front anti-roll bar stiffness is:

$$(k_t)_{ANR_f} = (k_t)_{vehivle} - \left(\frac{1}{2} k_s T_s^2\right)_f - \left(\frac{1}{2} k_s T_s^2\right)_r = 2323 \ Nm$$

Air Suspension System Design

4

The air suspension uses a combination of flexible air springs to transfer the chassis load to the axles. Originally, it was intended for heavy-duty applications, such as aircraft, trucks, buses, trains, etc. Nowadays, it is increasingly used in luxury cars, sport utility vehicles, and vans. Figure 4.1 shows a three-axle air suspension system developed by STAS for heavy-duty truck trailers.

Since 1920, French man George Messier started providing aftermarket pneumatic suspensions. In the 1950s, Citroen developed a four-wheel hydro-pneumatic suspension with self-leveling functionality that incorporated the features of earlier air suspensions. During the same period, General Motors equipped the 1957 Cadillac Eldorado Brougham with an air suspension, which compensated for uneven road conditions and maintain ride height. Meanwhile, Buick replaced the four conventional coil springs with pneumatic cylinders for suspension self-leveling, known as 'Air-Poised Suspension', by which the driver could control and raise the vehicle body by up to 5.5 inches. Based on this, Buick offered an optional 'Air Ride' suspension to customers which used steel coil springs in the front and air springs in the rear. Similarly, American Motors Corporation also offered the air suspension for Rambler Ambassadors at an additional cost of $99. However, it was not popular due to the cost and was discontinued after two years. In the 1960s, Borgward P100 introduced the first German self-leveling air suspension. After that, Mercedes Benz equipped their 300SE and 600 models with air suspensions. Besides, Rolls-Royce incorporated self-leveling air suspension on the 1965 Silver Shadow under the license from Citroen. In the 1970s, Mercedes Benz developed a hydro-pneumatic suspension and used it in the 450SEL model, which had a simpler structure with lower cost as shown in Fig. 4.2. In the 1980s, Toyota introduced the first electronically controlled air suspension, which contained adaptive dampers and air springs. Within the same period, the Electronically Controlled Air Suspension (ECAS) was trademarked by Dunlop Systems Coventry UK. This system was firstly applied to the Range Rover 93MY. In the 2000s,

A. Goodarzi et al., *Vehicle Suspension System Technology and Design*,
Synthesis Lectures on Advances in Automotive Technology,
https://doi.org/10.1007/978-3-031-21804-0_4

Fig. 4.1 Air suspension for truck trailers reproduced by permission of STAS

the GMT 360 Trail Blazer sport utility vehicle was equipped with an air suspension that used the integrated control system developed by WABCO and the air spring produced by Dunlop. Moreover, the Hummer H2 offered an optional rear air suspension with a dual air compressor which could also support tire inflation for off-road driving applications. Nowadays, most car factories offer various air suspension systems with more benefits than convolutional mechanical suspensions, such as lightweight, reduced vibration, improved ride quality, and adjustable ride heights under load variations (automatic and/or manual), etc.

Technically, the main components of an air suspension include air springs, air supply, and height-leveling system. The number and arrangement of the air springs are varied among applications. In general, there is at least one air spring for each corner of the suspension system. The air supply contains an air compressor, air tank, valves, and airlines. As the key mechanism of the air suspension, the height-leveling system dictates and adjusts the suspension height by controlling the air pressure which is also known as the suspension self-leveling system. The mechanical height-leveling system was firstly developed by the combination of mechanical leveling valves and linkages. Later on, the Electronic Controlled Air Suspension (ECAS) was invented, which used solenoid valves and height sensors to achieve better performance.

4.1 Air Spring

At the beginning, the air spring was designed for heavy aircraft. Nowadays, it is commonly used in the automotive industry to achieve the suspension self-leveling functionality and smooth driving quality. Different from leaf and the coil springs, the pressurized gas in the air spring provides the resistant forces to carry the load and attenuate vibrations. The air spring bellow is used to contain a certain amount of pressurized gas, which is usually made of rubber and fabric materials. During the 1930s, Roy Wilbur Brown developed the two-convolution air spring with fabric-reinforced rubber bellows, which became the first successful air spring that could be operated in harsh driving environments.

Fig. 4.2 Hydro-pneumatic
self-leveling suspension

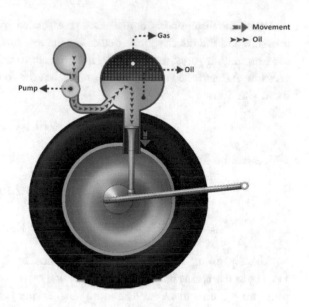

As shown in Fig. 4.3, the force exerted by the air spring equals the effective area multiplied by the internal air pressure as:

$$F = (P_S - P_A)A \tag{4.1}$$

in which A represents the effective area of the air spring that is non-constant during compression and extension, which equals to $\pi d^2/4$; P_A represents the atmospheric pressure; P_S represents the internal air pressure of the air spring. Since P_S is the dominant influence factor, the changes of the effective area A can be neglected around the equilibrium position for simplicity.

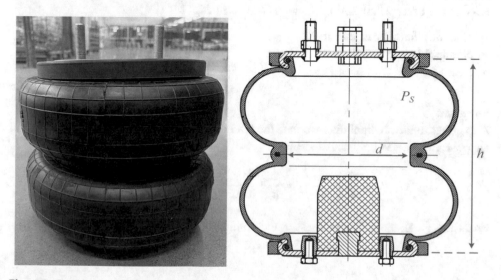

Fig. 4.3 Double convolution air spring

When the air spring is compressed or extended by a displacement Δx, the internal air pressure and the effective area determine the resistant force and the spring stiffness. With the nominal air pressure P_n and the nominal air volume V_n, the changes of the internal air pressure ΔP and the volume variation ΔV has the following relationship according to the Ideal Gas Law:

$$P_n V_n = (P_n + \Delta P)(V_n - \Delta V) \tag{4.2}$$

which can be linearized as

$$P_n V_n = P_n V_n - P_n \Delta V + \Delta P V_n - \underbrace{\Delta P \Delta V}_{\approx 0} \tag{4.3}$$

$$\Rightarrow P_n \Delta V = \Delta P V_n$$

The air spring stiffness K can be obtained by dividing the force variation ΔF with respect to the spring displacement Δx, which can be rewritten as a function of nominal air pressure P_n, effective area A, and nominal spring height h as

$$K = \frac{\Delta F}{\Delta x} = \frac{\Delta P A}{\Delta x} = \frac{\frac{P_n \Delta V}{V_n} A}{\Delta x} = \frac{\frac{P_n (A \Delta x)}{A h} A}{\Delta x} = \frac{P_n A}{h} \tag{4.4}$$

To maintain a constant height under various loads, the nominal air pressure needs to be adjusted to provide enough supporting force. Based on Eq. (4.4), it can be seen that the air spring becomes progressively stiffer under higher load. As a result, the air suspension's natural frequency can remain the same regardless of the chassis load. This is a great benefit for commercial vehicles whose loads are easily varied in tons.

Example 4.1: Based on the specifications given below, calculate the approximate stiffness of the example air spring model as shown in Fig. 4.3.

- Nominal diameter d: 160 mm
- Nominal height h: 240 mm
- Load: 500 kg.

Solution:
According to the nominal diameter and the load that carries by the air spring, the nominal air pressure P_n. can be calculated by

$$P_n = \frac{F}{A} = \frac{mg}{\pi d^2 / 4} = \frac{500 \times 9.81}{\pi \times 0.16^2 / 4} = 2.4359 \times 10^5 \ Pa$$

From Eq. (4.4), the spring stiffness at the nominal air pressure 1.249 bar is

$$K = \frac{P_n A}{h} = \frac{2.4359 \times 10^5 \times \pi \times 0.16^2/4}{0.24} = 20437\,N/m$$

It is noticeable that the air spring stiffness is highly dependent on the nominal air pressure. In order to carry a heavier sprung mass, the air spring needs to be charged at a higher pressure, which results in higher spring stiffness. Since the natural frequency is determined by the spring stiffness and the loaded mass, the air spring can maintain a relatively constant natural frequency that is beneficial to suspension performances.

Figure 4.4 describes the load deflection curve of the above air spring. It can be seen that the nominal air pressure P_n increases with the increase of the load F at the same height level. Specifically, the air spring needs to be charged at 3 bar to carry approximately 5.5 kN load, and 5 bar for nearly 10 kN load.

Example 4.2: Based on the requirements given below, determine the proper air spring size.

- Sprung mass natural frequency f_n: 1.4 Hz
- Load: 500 kg
- Nominal air pressure: 2.5 bar
- Nominal height h: 240 mm.

Fig. 4.4 Air spring load–deflection curve

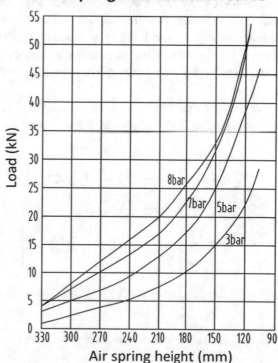

Air spring load-deflection curve

Solution:

According to the desired sprung mass natural frequency and the load, the spring stiffness can be calculated by

$$K = 4\pi^2 f_n^2 m = 4\pi^2 \times 1.4^2 \times 500 = 38689 \; N/m$$

Based on the nominal air pressure and the load, the nominal area of the air spring can be found as follows:

$$A = \frac{F}{P_n} = \frac{mg}{P_n} = \frac{500 \times 9.81}{2.5 \times 10^5} = 0.0196 \; m^2$$

Thus, the corresponding diameter is

$$d = \sqrt{\frac{4A}{\pi}} = \sqrt{\frac{4 \times 0.0196}{\pi}} = 0.158 \; m$$

4.2   Height-leveling System

There are different arrangements of the air suspension in practical applications. For a dual-axle air suspension system, there are normally three leveling valves used in the height-leveling system. The most common arrangement is one leveling valve for the front axle and the other two for the rear axle, as shown in Fig. 4.5a. Alternatively, some manufacturers use two leveling valves for the front axle and one for the rear axle, as shown in Fig. 4.5b. Besides, another arrangement shown in Fig. 4.5c is to use one leveling valve for each corner of the suspension, which is not common due to its cost and complexity.

In a mechanical height-leveling system, the leveling valves usually work with a linkage mechanism that is also known as the activating lever. As shown in Fig. 4.6, this L shape

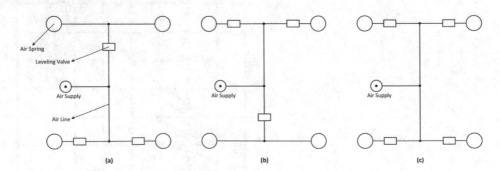

Fig. 4.5 Air suspension arrangements

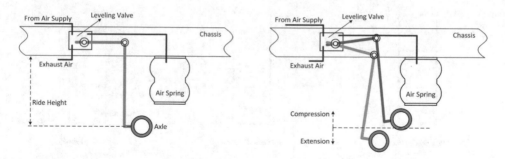

Fig. 4.6 Mechanical height-leveling system

linkage connects the axle to the leveling valve that is mounted on the chassis frame. When the axle moves up and down with respect to the chassis, the leveling valve reacts to the height changes, so that the air spring can be inflated or deflated accordingly. For example, when the chassis frame moves down, the linkage raises and the leveling valve connects the air supply to the air spring, which inflates the air spring and raises the chassis frame to the proper height. This process maintains a constant suspension height regardless of the chassis load.

In general, the leveling valve has three positions, i.e., inflate, neutral, and deflate positions. Figure 4.7 presents a conventional leveling valve introduced by STAS. When an extra load is added to the vehicle, the leveling valve is at the inflate position due to the movements of the linkage. After charging a certain amount of air into the air spring, the leveling valve switches to the neutral position when the desired ride height is achieved. Similarly, the leveling valve activates the deflate position when the load is removed from the vehicle. The air exists in the air spring through the exhaust port, then the leveling valve switches back to the neutral position when the chassis frame goes down to the proper height. Nowadays, some manufacturers start to use electronic leveling valves as shown in Fig. 4.8, which generally have the same positions as the conventional ones.

Moreover, there are two categories of mechanical leveling valves, i.e., the instant response valve and the delay valve. As the name describes, the instant response valve starts inflating or deflating the air springs as soon as the activating lever moves, while the delay valve responds with a slight delay before connecting the air springs to the air supply.

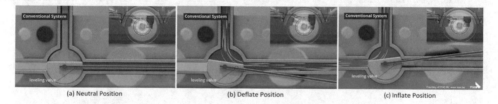

Fig. 4.7 A conventional leveling valve reproduced by permission of STAS

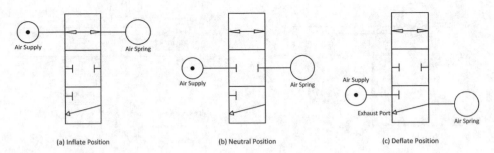

Fig. 4.8 Leveling valve

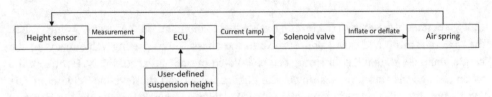

Fig. 4.9 The general operation procedures of ECAS

4.3   Electronic Controlled Air Suspension (ECAS)

The height-leveling system was firstly designed with mechanical leveling valves and linkages, but the Electronic Controlled Air Suspension (ECAS) was produced in the 1980s with the rapid development of electromechanical control systems. The use of electronic control unit (ECU) and height sensors improve the performance of the height-leveling system. Besides, it is possible to remotely control the ride height which allows the user to adjust the level easily. The general operating procedures of the ECAS is presented in Fig. 4.9, which can be concluded as follows:

(1) A height sensor is installed in parallel with the air spring, which measures the distance between the chassis frame and the axle. The measurement is then sent to the ECU for processing.
(2) The ECU compares the measured signal to the desired suspension height that is determined by the user. According to their difference, a control signal is sent to the solenoid valve that controls the inflating and deflating of the air spring as a leveling valve.
(3) With the control of the solenoid valve, the pressure inside the air spring changes accordingly to adjust the suspension height.
(4) The updated distance between the chassis frame and the axle is measured by the height sensor again, and the control cycle goes back to Step 1.

Adaptive Suspension Systems Design

5

The vehicle suspension system aims to provide a comfortable ride and stable handling. However, these two objectives are conflicting, so the conventional passive suspension design has to compromise between ride comfort and handling stability. Thus, active and semi-active suspension systems are developed to further improve suspension performances.

5.1 Active Suspensions

Active suspensions are fundamentally different from traditional suspension systems. A traditional suspension stores energy in a spring and dissipates it via a shock absorber whose damping characteristics are fixed. An active suspension is also able to store and dissipate energy, but the passive shock absorbers are replaced with high-speed actuators. Figure 5.1 shows the use of a fully active suspension system in a car.

The active suspension system can handle bumps and rough roads much better by adjusting the actuator forces in real time. Fully active systems can control body motions (pitch, roll, and bounce) as well as wheel hub motion over a wide range of frequencies (from 0 to 30 Hz). Its high cost, high power consumption, and complexity are the known disadvantages of fully active suspensions. From a control strategy point of view, active systems could use preview control methods that can be divided into two main categories: the wheelbase preview method that can detect wheel conditions, and the look-ahead preview method that can predict future road conditions.

During the 1980s, few vehicles used the active suspension technology, which were mainly test vehicles and most notably from Lotus and Volvo. During the 1990s, interest in active systems declined due to their cost, complication, and high power consumption.

A. Goodarzi et al., *Vehicle Suspension System Technology and Design*, Synthesis Lectures on Advances in Automotive Technology, https://doi.org/10.1007/978-3-031-21804-0_5

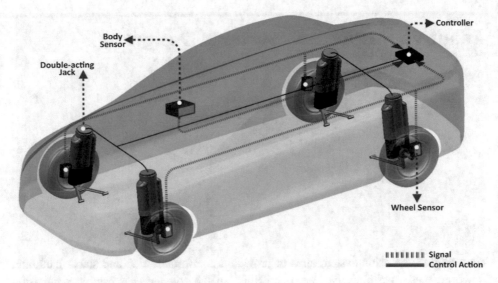

Fig. 5.1 Active suspension system components

There are a few partially active systems on the market, but a fully active system is a combination of a passive system for the high-frequency range and an active system for the low-frequency range.

5.2 Semi-active Suspensions

To reduce the complicity and cost of fully active suspensions, the semi-active suspensions received a lot of attention in the automobile industry. Nowadays, they are widely used in high-end cars. The main working principle of a semi-active system is the damping control that provides a distinct advantage in both ride quality and handling. Figure 5.2 shows the main parts of a semi-active suspension. This system includes adaptive shock absorbers as its main part, multiple sensors such as height sensors and IMU to detect road conditions and vehicle motions, and an Electronic Control Unit (ECU) that controls the damping properties of the adaptive shock absorbers to improve vehicle comfort and/or handling stability. In normal driving, the ECU reduces the damping coefficient to provide ride comfort; however, in cornering maneuvers, the damping coefficient is increased to improve vehicle handling and safety. A semi-active suspension can only optimize body motions (bump, pitch, or roll) at a low-frequency range (from 1 to 4 Hz); whereas the high-frequency range is handled by the passive suspension setting.

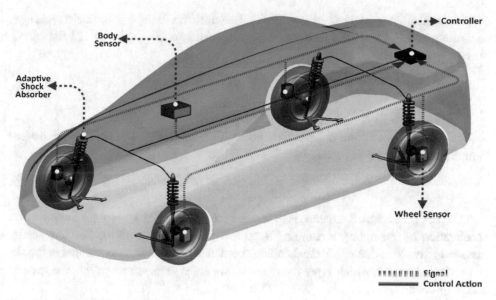

Fig. 5.2 Semi-active suspension system

5.3 Skyhook Control

In the early 1970s, Karnopp and Crosby introduced the concept of the well-known Skyhook control. Because of its intuitive approach and good performance properties, it is now widely used in vehicle semi-active suspension systems. The concept of the Skyhook approach is that the sprung mass connects to an imaginary fixed sky through a shock absorber, and a spring is installed between the sprung and the unsprung mass, as shown in Fig. 5.3.

Fig. 5.3 The schematic diagram of the Skyhook control approach

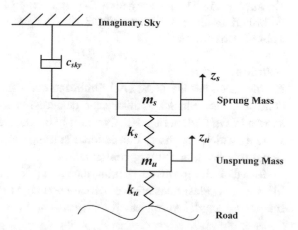

This control law consists of two states, in which the damping coefficient changes according to the sign of the product of the suspension relative velocity and the sprung mass velocity as.

$$
\begin{cases}
c_{sky} = c_{sky-max}, & if \ \dot{z}_s(\dot{z}_s - \dot{z}_u) \geq 0 \\
c_{sky} = c_{sky-min}, & if \ \dot{z}_s(\dot{z}_s - \dot{z}_u) < 0
\end{cases}
\tag{5.1}
$$

The damping coefficient c_{sky} switches beeen high-state and low-state, then the desired damping force can be calculated by.

$$
F_{sky} = -c_{sky}\dot{z}_s \tag{5.2}
$$

This comfort-oriented control law focuses on reducing the absolute sprung mass acceleration by generating a damping force that is proportional to the absolute sprung mass velocity. Note that c_{sky} is the damping coefficient of the imaginary damper in the sky as shown in Fig. 5.2, which is not the realistic one in the suspension system. The realistic damping coefficient c_{damper} can be calculated by

$$
c_{damper} = -\frac{F_{sky}}{\dot{z}_s - \dot{z}_u} \tag{5.3}
$$

It is worth mentioning that the above damping coefficient is usually constrained because of the mechanical limitations of the adaptive shock absorber as

$$
c_{damper-min} \leq c_{damper} \leq c_{damper-max} \tag{5.4}
$$

Then, the realistic damping force can be found as

$$
F_{damper} = c_{damper}(\dot{z}_u - \dot{z}_s) \tag{5.5}
$$

Example 5.1: Design an active Skyhook controller for a 1-DOF quarter car system described in Sect. 3.5.1 and Fig. 3.19. The relevant vehicle parameters are listed in Table 5.1 as follows:

Solution
In this example, the MATLAB/SimScape software is utilized to evaluate the Skyhook controller. The SimScape library can be accessed from the Simulink Library Browser, as shown in Fig. 5.4. In Fig. 5.5, five built-in blocks are used to set up a mass-spring-damper system, which are mass, translational spring, translational damper, mechanical translational reference, and solver configuration.

Based on the above mass-spring-damper system, a disturbance input is added through a block called ideal translational velocity source as shown in Fig. 5.6. This is the 1-DOF quarter-car model with a passive suspension.

Table 5.1 Specifications of a quarter car model

Symbol	Parameter	Value
m_s	Sprung mass	400 kg
k_s	Spring stiffness	28 kN/m
c_p	Passive damping coefficient	2500 Ns/m
$c_{sky-max}$	Maximum damping coefficient used in Skyhook algorithm	5000 Ns/m
$c_{sky-min}$	Minimum damping coefficient used in Skyhook algorithm	500 Ns/m
$c_{damper-max}$	Maximum damping coefficient of the adaptive shock absorber	1000 Ns/m
$c_{damper-min}$	Minimum damping coefficient of the adaptive shock absorber	4000 Ns/m

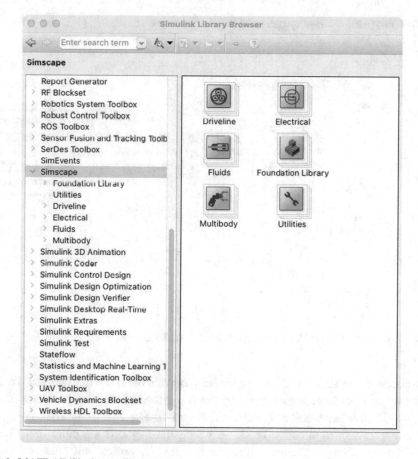

Fig. 5.4 MATLAB/SimScape library

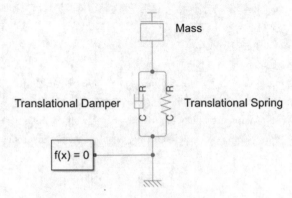

Fig. 5.5 Mass-spring-damper system in MATLAB/SimScape

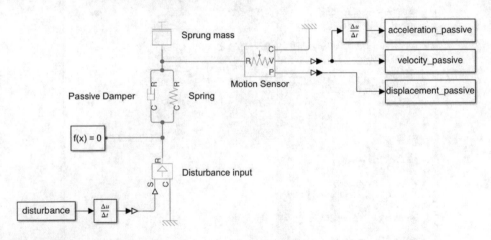

Fig. 5.6 Passive 1-DOF quarter-car model in MATLAB/SimScape

Compared to the above 1-DOF quarter-car passive model, the translational damper is replaced by an ideal force source block which is operated as the adaptive shock absorber in the semi-active suspension system, as shown in Fig. 5.7. It is connected to the MATLAB user-defined function block, in which the Skyhook algorithm introduced in Eqs. (5.1) and (5.2) is programmed. It can be seen that the vertical velocities of the sprung mass zs_dot and the disturbance input zu_dot are feedback to the Skyhook algorithm for calculating the control force. The MATLAB code used in this example is as follows:

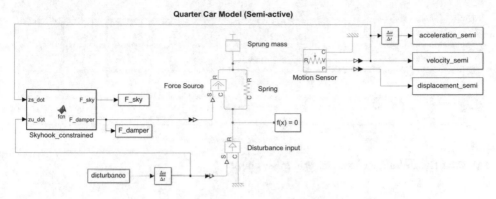

Fig. 5.7 Semi-active 1-DOF quarter-car model in MATLAB/SimScape

```
c_sky_max = 5000;
c_sky_min = 500;
delta_v = zs_dot - zu_dot; % suspension relative velocity
if delta_v * zs_dot >= 0    % high state
        c_sky = c_sky_max;
    else   % low state
        c_sky = c_sky_min;
end
F_sky = - c_sky * zs_dot;
```

After that, a *if* loop is utilized to implement the damping constraint mentioned in Eqs. (5.3) and (5.4):

```
c_damper = - F_sky/(delta_v+eps);
c_damper_max = 4000;
c_damper_min = 1000;
if c_damper > c_damper_max
    c_damper = c_damper_max;
elseif c_damper < c_damper_min
    c_damper = c_damper_min;
else
    c_damper = c_damper;
end
F_damper = - c_damper * delta_v;
```

In MATLAB, *eps* refers to the floating-point relative accuracy. Its value depends on the data type used in the code. For example, if the data type is "double", the value of *eps* is 2^{-52}, and it equals to 2^{-23} when the data type is "single". In this example, *eps* is used to prevent a zero denominator when the suspension relative velocity *delta_v* equals to 0. Then, the damping coefficient is limited within the possible range from *c_damper_min* to *c_-damper_max*. In the end, the realistic damping force *F_damper* that can be generated by the adaptive shock absorber is calculated according to Eq. (5.5).

The disturbance input used in this example is shown in Fig. 5.8, which simulates that a vehicle is driven over a speedbump at 20 km/h. Based on the selected parameters, the displacement, velocity, and acceleration of the sprung mass are plotted in Figs. 5.9, 5.10 and 5.11, respectively. The blue curve represents the responses of the passive suspension, and the red curve represents the semi-active suspension with a Skyhook controller. It is

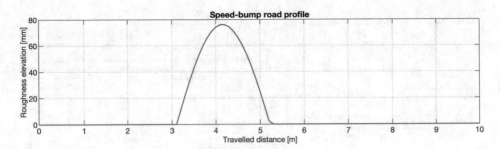

Fig. 5.8 The disturbance input used in the example

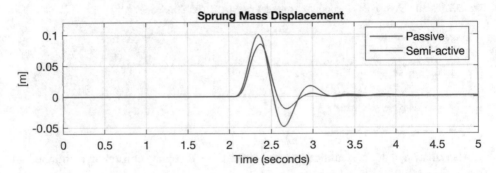

Fig. 5.9 The displacement responses of sprung mass (passive and semi-active)

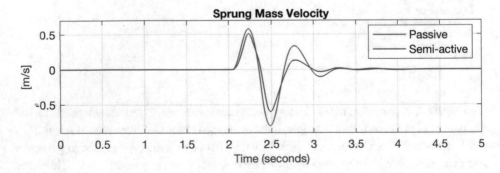

Fig. 5.10 The velocity responses of sprung mass (passive and semi-active)

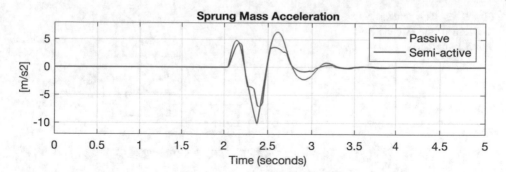

Fig. 5.11 The acceleration responses of sprung mass (passive and semi-active)

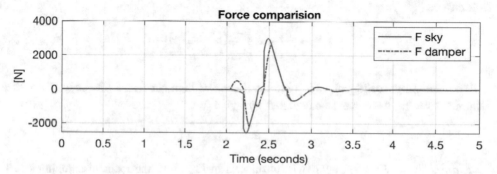

Fig. 5.12 The comparison between the desired control force and the constraint force

noticeable that the vibrations are attenuated to a large extent after implementing the Skyhook controller.

The calculated Skyhook force F_sky and constrained damping force F_damper are presented and compared in Fig. 5.12. It can be seen that sometimes the force requested by the Skyhook controller cannot be achieved due to the damping dissipative constraint and limitations.

5.4 Groundhook Control

As we discussed earlier, the Skyhook approach only dissipates energy from the sprung mass, but the motion of the unsprung mass could become excessive that makes the vehicle handling worse. To overcome this issue, the Groundhook configuration is developed by moving the Skyhook damper to the unsprung mass. As can be seen in Fig. 5.13, the imaginary Groundhook damper is connected between the fixed ground and the unsprung mass.

Fig. 5.13 The schematic
diagram of the Groundhook
control approach

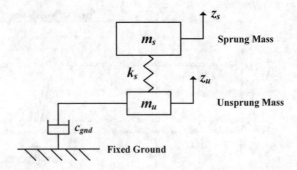

This control law also consists of two states, in which the damping coefficient changes according to the sign of the product of the suspension relative velocity and unsprung mass velocity as

$$\begin{cases} c_{gnd} = c_{max}, \ if \ \dot{z}_u(\dot{z}_s - \dot{z}_u) \leq 0 \\ c_{gnd} = c_{min}, \ if \ \dot{z}_u(\dot{z}_s - \dot{z}_u) > 0 \end{cases} \qquad (5.6)$$

The damping coefficient c_{gnd} switches between high-state and low-state, then the desired damping force can be calculated by

$$F_{gnd} = -c_{gnd}\dot{z}_u \qquad (5.7)$$

Similar to the Skyhook approach introduced in Eq. (5.3), the realistic damping coefficient c_{damper} can be calculated by

$$c_{damper} = -\frac{F_{gnd}}{\dot{z}_s - \dot{z}_u} \qquad (5.8)$$

Furthermore, the damping constraint also needs be considered in a semi-active Groundhook controller as introduced in Eq. (5.4).

Printed in the United States
by Baker & Taylor Publisher Services